Managing Reality
Book Four
Managing Change

Third edition

Other titles in the Managing Reality series:

Managing Reality. Book One: Introduction to the Engineering and Construction Contract. Third edition (2017)
Bronwyn Mitchell and Barry Trebes. ISBN 978-0-7277-6182-8

Managing Reality. Book Two: Procuring an Engineering and Construction Contract. Third edition (2017)
B. Mitchell and B. Trebes. ISBN 978-0-7277-6184-2

Managing Reality. Book Three: Managing the Contract. Third edition (2017)
B. Mitchell and B. Trebes. ISBN 978-0-7277-6186-6

Managing Reality. Book Five: Managing Procedures. Third edition (2017)
B. Mitchell and B. Trebes. ISBN 978-0-7277-6190-3

Managing Reality
Book Four
Managing Change

Third edition

Bronwyn Mitchell and Barry Trebes

Published by ICE Publishing, One Great George Street, Westminster, London SW1P 3AA.

Full detail of ICE Publishing representatives and distributors can be found at:
www.icebookshop.com/bookshop_contact.asp

Other titles by ICE Publishing:

NEC4: The Role of the Project Manager
B. Mitchell and B. Trebes. ISBN 978-0-7277-6353-2
NEC4: The Role of the Supervisor
B. Mitchell and B. Trebes. ISBN 978-0-7277-6355-6
NEC3 and NEC4 Compared
R. Gerrard. ISBN 978-0-7277-6201-6

www.icebookshop.com

A catalogue record for this book is available from the British Library

ISBN 978-0-7277-6188-0

© Thomas Telford Limited 2018

ICE Publishing is a division of Thomas Telford Ltd, a wholly-owned subsidiary of the Institution of Civil Engineers (ICE).

All rights, including translation, reserved. Except as permitted by the Copyright, Designs and Patents Act 1988, no part of this publication may be reproduced, stored in a retrieval system or transmitted in any form or by any means, electronic, mechanical, photocopying or otherwise, without the prior written permission of the Publisher, ICE Publishing, One Great George Street, Westminster, London SW1P 3AA.

This book is published on the understanding that the authors are solely responsible for the statements made and opinions expressed in it and that its publication does not necessarily imply that such statements and/or opinions are or reflect the views or opinions of the publishers. While every effort has been made to ensure that the statements made and the opinions expressed in this publication provide a safe and accurate guide, no liability or responsibility can be accepted in this respect by the author or publishers.

While every reasonable effort has been undertaken by the author and the publisher to acknowledge copyright on material reproduced, if there has been an oversight please contact the publisher and we will endeavour to correct this upon a reprint.

Commissioning Editor: Michael Fenton
Production Editor: Madhubanti Bhattacharyya
Market Development Executive: Elizabeth Hobson

Typeset by Academic + Technical, Bristol
Index created by Laurence Errington
Printed and bound by NOVOPRINT S.A., Spain

Contents

	Preface	vii
	Foreword	ix
	Acknowledgements	xi
	Series contents	xiii
	Introduction	**1**
	General	1
	Background	1
	The structure of the books	2

01 — **Compensation events** — **3**

	Synopsis	3
1.1.	Introduction	4
1.2.	Compensation event procedure – background	4
1.3.	What is a compensation event?	5
1.4.	Where to find a list of compensation events	6
1.5.	Roles of the *Project Manager* and *Contractor*	18
1.6.	Administering compensation events	18
1.7	Notification of a compensation event	20
1.8.	Quotations for a compensation event	23
1.9.	Assessment of quotations	26
1.10.	The use of the programme for the assessment of compensation events	27
1.11.	Implementation of compensation events	28
1.12.	Reduction of Prices	28
1.13	Frequently asked questions	29
1.14.	Format of a compensation event quotation	33

Appendix 1 — **Compensation event procedure** — **35**

	Section A: Actions required to be taken by the *Project Manager*	36
	Section B: Actions required to be taken by the *Contractor*	37
	Section C: Compensation event procedure	38
	Section D: Procedure for events notified by the *Project Manager*	40
	Section E: Procedure for events notified by the *Contractor*	41
	Section F: Most common complicating factors	42

02 — **Schedule of Cost Components and Short Schedule of Cost Components** — **43**

	Synopsis	43
2.1.	Introduction	44
2.2.	What is the SCC?	44
2.3.	Why has this approach been taken?	45
2.4.	Assessment options	45
2.5.	When is the SCC or SSCC used?	45
2.6.	Defined Cost	48
2.7.	The Fee	49
2.8.	The components of cost included under the SCC	49
2.9.	The components of cost included under the SSCC	56
2.10.	Contract Data part two	60
2.11.	Putting it all together for payment – Option C	60
2.12.	Inspecting accounts and records	63
2.13.	Use of the SSCC	64
2.14.	Practical issues	64
2.15.	Preliminaries and people costs	70

Appendix 2 — **Example quotations for compensation events** — **71**

	Section A: Based on the SCC	71
A2.1.	Introduction	71
A2.2.	Contract Data example	73
A2.3.	Defined Cost	74
A2.4.	Direct fee percentages	74
A2.5.	SCC example quotation	75

	A2.6. Supporting notes	87
	Section B: Based on the SSCC	95
	B2.1. Introduction	95
	Section C: Based on rates or lump sums – all main Options	97
	C3.1. Introduction	97
Appendix 3	**Example people cost calculations**	**99**
	A3.1. Introduction	99
Appendix 4	**Comparison between traditional preliminaries build-up and how they relate to the Schedule of Cost Components and the Short Schedule of Cost Components**	**101**
	A4.1. Introduction	101
Appendix 5	**Interrelationship between the *Contractor's* and the Subcontractor's *share* on target cost contracts**	**107**
	A5.1. Introduction	107
	Index	**109**

Preface

In the preface to the first edition of *Managing Reality*, in 2005, we set out our aims and aspirations for 'managing reality'. These were as follows:

- to add and contribute to the body of knowledge on the use of the NEC Engineering and Construction Contract (ECC)
- to provide a set of books that focuses on the 'how to' – how to manage and administer the ECC contract
- to present as a five-part book series that covers both the needs of the student professional or prospective client, through to the novice practitioner and experienced user
- to provide a rounded view of the ECC, whatever your discipline, on both sides of the contractual relationship
- to enable everyone to realise the business benefits from using the NEC suite of contracts generally and the ECC in particular.

Managing Reality does not attempt to give a legal treatise or a blow-by-blow review of each and every clause. It is intended to be complementary to other publications, which give excellent theoretical and legal perspectives.

This book is about dealing with the reality of real life projects: managing reality.

The feedback and support we have received since the first publication of *Managing Reality* in 2005 has been universally positive, and we would like to thank all of you who have bought and used it since its first publication.

We have greatly enjoyed updating and working on this third edition, and we hope that these books continue to provide a useful body of knowledge on the use of the NEC4 ECC.

Bronwyn Mitchell and Barry Trebes
2017

Foreword

A key objective of the first edition of *Managing Reality* was to provide a five-part book series to meet the needs of students, prospective clients, novice practitioners and experienced users. Satisfying such diverse needs is an ambitious objective for any text.

Does *Managing Reality* achieve its stated aim? I believe that the answer to this is a resounding 'yes'. In my view, the calibre of authorship is exceptional. All levels of and types of readership from the uninitiated to the experienced professional will derive considerable benefit from this text. Although written in a very accessible style, there is no skimping on detail or on addressing difficult issues. The worked examples are particularly helpful. *Managing Reality* should be your prime aid from the moment you are considering whether or not to use an NEC contract right through to using and operating the contract.

But *Managing Reality* is much more than simply a 'how to' guide. It seeks to deliver a clear message that NEC contracts cannot be used to their full potential unless one is prepared to ditch one's knowledge and experience of traditional contracting. For example, emphasis is placed on the fact that certainty and predictability are the hallmarks of NEC contracts. Open-ended and subjective phrases and concepts have no place in NEC contracting.

I am privileged to be associated with this third edition of *Managing Reality*. Its publication is very timely since it coincides with the publication of the NEC4 suite of contracts. This new edition of the NEC suite is a reflection of the dominant position that NEC contracts now have both in the UK and in many other countries. It will continue to support those who need help overcoming any reservations about using NEC contracts, and reinforce existing users in their continued use of these ground-breaking contracts.

Professor Rudi Klein
President, NEC Users' Group

Acknowledgements

We would like to thank the following individuals and companies who have supported this book.

For their active participation in this book we would like to thank

- Professor Rudi Klein (SEC Group Chief Executive) for writing the Foreword
- Dr Robert N. Hunter of Hunter and Edgar Edinburgh for his thoughts and suggested revisions for this third edition
- Michael Fenton, the ICE commissioning editor, for his enthusiasm and patience
- Richard Patterson of Mott MacDonald
- everyone who has given feedback on this book since 2005.

And our continued gratitude to those who provided support and input into the first edition of *Managing Reality*:

- Mike Attridge, of Ellenbrook Consulting, who reviewed this book on behalf of the authors
- David H. Williams, who provided guidance and support in the development of the first edition of this book
- everyone at Needlemans Limited Construction Consultants (now part of the Mott MacDonald Group)
- everyone at MPS Limited with whom Needlemans Limited worked to develop the first web-based management system for the NEC in 2000.

Finally, we would like to thank our family and friends for their ongoing support, understanding and patience.

Series contents

The following outlines the content of the five books in the series.

Book 1   Managing Reality: Introduction to the Engineering and Construction Contract

Chapter 1   **Introduction to the Engineering and Construction Contract, concepts and terminology**
This chapter looks at

- an introduction to the ECC
- an identification of some of the differences between the ECC and other contracts
- an outline of the key features of the ECC
- conventions of the ECC
- concepts on which the ECC is based
- terminology used in the ECC
- terminology not used in the ECC
- how the ECC affects the way you work.

Chapter 2   **Roles in the Engineering and Construction Contract**
This chapter describes the roles adopted in the ECC, including

- how to designate a role
- discussion of the roles described in the ECC
- discussion of the project team
- how the ECC affects each of the roles.

Appendix 1   **List of duties**

Book 2   Managing Reality: Procuring an Engineering and Construction Contract

Chapter 1   **Procurement**
This chapter looks at the concept of procurement and contracting strategies and discusses

- procurement and contract strategy
- what tender documents to include in an ECC invitation to tender
- how to draft and compile a contract using the ECC
- procurement scenarios that a client could face and how to approach them
- what are framework agreements and how they could incorporate the ECC
- what is partnering and how it can be used with the ECC.

Appendix 1   **Assessing tenders**

Appendix 2   **ECC tender documentation**

Chapter 2   **Contract Options**
This chapter looks at the Contract Options available within the ECC, including

- ECC main and secondary Options
- priced contracts
- target contracts
- cost-reimbursable contracts
- choosing a main Option
- choosing a secondary Option.

Appendix 3	Inspection of the *Contractor's* accounts and records plan
Chapter 3	**Completing the Contract Data**

This chapter gives guidance on

- how to choose a main Option
- how to choose an Option for resolving and avoiding disputes
- how to choose secondary Options
- choosing optional statements in the Contract Data
- where to position the optional statements in the Contract Data
- how to complete each statement in the Contract Data.

Chapter 4	**Scope guidelines**

This chapter looks at the Scope and Site Information, including

- Providing the Works
- what should be included in the Scope
- separation of the Scope and the Site Information
- structuring for the Scope
- interface management
- general rules in drafting the Scope
- Site Information.

Appendix 4	**Scope clauses**

Book 3	Managing Reality: Managing the Contract

Chapter 1	**Payment procedures in the Engineering and Construction Contract**

This chapter discusses the following

- the payment procedure, including
  - when the *Contractor's* application for payment is submitted
  - when assessments take place
  - when the payment certificate is issued
  - how invoicing is carried out
  - when payment takes place
- the effects of option Y(UK)2, taking into account the Housing Grants, Construction and Regeneration Act 1996 as amended.

Chapter 2	**Control of time**

This chapter discusses aspects relating to the *Contractor's* programme, including

- the terminology used to describe the programme
- what the programme is
- the definition and purpose of the Accepted Programme
- how and when to submit programmes
- what the programme is used for
- what to include in the programme.

Chapter 3	**Control of quality**

This chapter discusses:

- The quality framework embedded within the ECC
- The *Contractor*'s obligations
- Role of the *Employer*'s representatives
- Subcontracting
- Quality control

Chapter 4	**Disputes and dispute resolution**

This chapter

- emphasises the importance of early dispute resolution to the successful outcome of a contract
- considers the common sources of dispute
- considers how the ECC has been designed to reduce the incidence of disputes
- examines how the ECC provides for the resolution of disputes
- looks at the implications for the dispute resolution process as a result of the HGCR Act as amended
- looks at ECC changes in relation to adjudication.

Book 4	**Managing Reality: Managing Change**
Chapter 1	**Compensation events**

This chapter describes

- the compensation events contained within the ECC
- procedure for administering compensation events
- roles played by the two main parties to the contract in relation to compensation events.

Appendix 1	**Compensation event procedure**
Chapter 2	**Schedule of Cost Components and Short Schedule of Cost Components**

This chapter discusses aspects relating to the full SCC and its short version, the SSCC, including

- when the SCC and SSCC are used
- how the SCC and SSCC interact with the payment clauses
- Defined Cost
- the Fee
- the components of cost included under the SCC and SSCC
- Contract Data part two.

Appendix 2	**Example quotations for compensation events**
Appendix 3	**Example people cost calculations**
Appendix 4	**Comparison between traditional preliminaries build-up and how they relate to the Schedule of Cost Components and the Short Schedule of Cost Components**
Appendix 5	**Interrelationship between the *Contractor's* and the Subcontractor's *share* on target cost contracts**
Book 5	**Managing Reality: Managing Procedures**
Chapter 1	**ECC Management: Procedures**

This chapter brings together all the aspects discussed in previous chapters in Books One to Four of the *Managing Reality* series. This chapter provides the 'how to' part of the series. It introduces some example pro formas for use with the ECC. Unless detailed separately due to a complex procedure, replies are described under the relevant action.

For quick reference, this chapter may be read on its own. It does not, however, detail the reasons for carrying out the actions, or the clause numbers that should be referred to in order to verify the actions in accordance with the contract. These are described in detail in the other books that form part of this series.

Managing Change
ISBN 978-0-7277-6188-0

ICE Publishing: All rights reserved
http://dx.doi.org/10.1680/mc.61880.001

Introduction

General

This series of books will provide the people who are actually using the Engineering and Construction Contract (ECC) in particular, and the New Engineering Contract (NEC) suite in general, practical guidance as to how to prepare and manage an ECC contract with confidence and knowledge of the effects of their actions on the contract and the Parties.

Each book in the series addresses a different area of the management of an ECC contract:

- Book One – *Managing Reality: Introduction to the Engineering and Construction Contract*
- Book Two – *Managing Reality: Procuring an Engineering and Construction Contract*
- Book Three – *Managing Reality: Managing the Contract*
- Book Four – *Managing Reality: Managing Change*
- Book Five – *Managing Reality: Managing Procedures*.

Book One (*Managing Reality: Introduction to the Engineering and Construction Contract*) is for those who are considering using the ECC but need further information, or those who are already using the ECC but need further insight into its rationale. It therefore focuses on the fundamental cultural changes and mind-shift that are required to successfully manage the practicalities of the ECC in use.

Book Two (*Managing Reality: Procuring an Engineering and Construction Contract*) is for those who need to know how to procure an ECC contract. It covers in practical detail the invitations to tender, evaluation of submissions, which option to select, how to complete the Contract Data and how to prepare the Scope. The use of this guidance is appropriate for clients, contractors (including subcontractors) and construction professionals generally.

Book Three (*Managing Reality: Managing the Contract*) is essentially for those who use the contract on a daily basis, covering the detail of practical management such as paying the contractor, reviewing the programme, ensuring the quality of the *works*, and dispute resolution. Both first-time and experienced practitioners will benefit from this book.

Book Four (*Managing Reality: Managing Change*) is for those who are managing change under the contract; whether for the client or the contractor (or subcontractor), the management of change is often a major challenge whatever the form of contract. The ECC deals with change in a different way to other, more traditional forms. This book sets out the steps to efficiently and effectively manage change, bridging the gap between theory and practice.

Book Five (*Managing Reality: Managing Procedures*) gives step-by-step guidance on how to apply the most commonly-used procedures, detailing the actions needed by all Parties to comply with the contract. Anyone administering the contract will benefit from this book.

Background

The ECC could be termed a 'modern contract' in that it seeks to holistically align the setting up of a contract to match business needs, as opposed to writing a contract that merely administers construction events.

The whole ethos of the ECC, and the NEC suite generally, is one of simplicity of language and clarity of requirement. It is important that the roles and responsibilities of all of those involved in the contract are equally clear in definition and ownership.

When looking at the ECC for the first time it is easy to believe that it is relatively straightforward and simple. However, this apparent simplicity belies the need for the people involved to think about their project and their role, and how the ECC can deliver their particular contract strategy.

The ECC provides a structured flexible framework for setting up an appropriate form of contract whatever the selected procurement route. The fundamental requirements are as follows:

- The Scope – quality and completeness: what are you asking the *Contractor* to do?
- The Site Information: what are the site conditions that the *Contractor* will find?
- The Contract Data – key objectives for completion (e.g. start date, completion date and programme): when do you want it completed?

The details contained in this series of books will underline the relevance and importance of the above three fundamental requirements.

The structure of the books

Each chapter starts with a synopsis of what is included in that chapter. Throughout the book there are shaded 'practical tip' boxes that immediately point the user towards important reminders for using the ECC (see example below).

> Clarity and completeness of the Scope is fundamental.

There are also unshaded boxes that contain examples to illustrate the text (see example below).

> Imagine a situation in which the *Supervisor* notifies the *Contractor* that the reinstatement of carriageways on a utility diversion project is not to the highway authority's usual standards. However, the Scope is silent about the reinstatement.
>
> The test of a Defect is also whether the work is in accordance with the applicable law. In this instance, the reinstatement is not in accordance with the Road and Street Works Act 1991.

Other diagrams and tables are designed to maintain interest and provide another medium of explanation. There are also standard forms for use in the administration and management of the contract, together with examples.

Throughout the books, the following terms have been used in a specific way:

- NEC is the abbreviation for the suite of New Engineering Contracts and it is not the name of any single contract
- ECC is the abbreviation for the contract in the NEC suite called the Engineering and Construction Contract.

The NEC4 suite comprises the

- Engineering and Construction Contract
- Engineering and Construction Subcontract
- Engineering and Construction Short Contract
- Engineering and Construction Short Subcontract
- Professional Service Contract
- Professional Service Short Contract
- Term Service Contract
- Term Service Short Contract
- Supply Contract
- Supply Short Contract
- Framework Contract
- Dispute Resolution Service Contract
- Design Build and Operate Contract
- Alliance Contract – consultative version.

Managing Change
ISBN 978-0-7277-6188-0

ICE Publishing: All rights reserved
http://dx.doi.org/10.1680/mc.61880.003

Chapter 1
Compensation events

Synopsis   This chapter describes

- the compensation events contained within the ECC
- procedure for administering compensation events
- roles played by the two main parties to the contract in relation to compensation events.

1.1. Introduction

One of the underlying principles of the ECC is to avoid and reduce the amount of change that occurs on construction projects. However, the contract recognises that change is inevitable even when the project has been well planned and prepared, and it sets out to deal with the effects and consequences of change in an improved way.

The contract recognises that the earlier an event that could affect the cost, time or quality of a project is identified, then the more likely it is that its effects can be reduced or even avoided. The ECC early warning procedure is what it says: an early warning mechanism for change. Should the change become necessary, then the contract tries to deal with the effects of change in an improved way by use of quotations for compensation events.

Compensation events represent the mechanism for the *Contractor*, ensuring that it is not out of pocket for things that happen that are outwith its control. Unlike some traditional contracts, which address 'extensions of time' and 'variations' separately, the ECC regards changes as a package of time and money. This means that for every event the effects on both the programme (the Completion Date and Key Dates) and the contract sum (the total of the Prices) are considered at the same time.

Apart from a clear and finite list of events that could trigger a compensation event, other points to note are that

- the *Contractor* can notify a compensation event if it is less than 8 weeks since it became aware of the event (except in specific circumstances)
- a weather event is not confined to an extension of time assessment
- physical conditions, such as ground conditions, rely on information provided by the *Client*
- the procedure for most compensation events takes place within a maximum time period of 6 weeks.

> Compensation events are events that are at the *Client's* risk in the contract.

1.2. Compensation event procedure – background

In order to understand why the compensation event procedure has developed into the form found in the ECC, it is necessary to consider some of the principles upon which the contract is founded.

Every procedure has been designed so that its implementation should contribute to, rather than detract from, the effectiveness of the management of the work. In this context, management includes cost and time management.

The ECC is based on the principle that foresight applied collaboratively mitigates problems and shrinks risk. This could be considered a departure from traditional contracts that tend to view the Engineer/Architect/Supervising Officer as the font of all knowledge, paying little or no regard to any worthwhile contribution that the *Contractor* may have to offer in the area of problem resolution. Latham acknowledged this issue in his report, *Constructing the Team*, when advocating that a modern form of contract should include 'firm duties of teamwork with shared financial motivation to pursue the teamwork approach. These should involve a general presumption to achieve "win–win" solutions to problems that may arise during the course of a project.' Continuing on this theme, the ECC motivates people to play their part in collaborative management if it is in their commercial (*Client* and *Contractor*) and professional (consultants, e.g. designers and *Project Managers*) interest to do so.

The ECC sets out to motivate these people by clearly defining the actions to be taken, clearly stating who is responsible for taking those actions, and giving periods within which the actions are required to be taken. Such is the emphasis the ECC places on collaborative management that sanctions exist within the contract to be applied to the party who does not play their part. These sanctions take the form of financial consequences (direct and indirect) for such lapses as

- failure to reply within the *period for reply*
- failure to give early warnings
- failure to keep the programme up to date
- failure to submit quotations on time.

Another principle to assist with the efficient management of the *works* is that the *Project Manager*, acting on behalf of the *Client* and in communication with it, should be presented with options for dealing with a 'problem'. The *Contractor* should be indifferent to the choice made in terms of time and money.

This is achieved by basing the valuation of compensation events on a forecast of their impact upon the cost to the *Contractor* of carrying out the *works* as forecast by them at the time the event is assessed. Where, as is often the case, alternative ways of dealing with the 'problem' are possible, the *Contractor* prepares quotations for different ways of tackling it. Clause 62.1 requires the practicable options to be discussed with the *Contractor*. The *Project Manager* selects the ones which will serve the best interests of the *Client*. In some cases this will be the lowest cost solution, in others it might be the least-delay solution, or a combination of factors.

The financial effects of a compensation event are based upon a quotation prepared by the *Contractor*, preferably in advance of the work (the subject of the compensation event) being carried out. Under price-based contracts (main Options A and B), the *Contractor* carries the risk if its forecast of financial effect turns out to be wrong and consequently the *Client* has a firm commitment. Under the target cost contracts (main Options C and D), the *Contractor* carries some risk if its forecast is wrong, as it will affect their final 'share' from the target mechanism. This is justified on the grounds that

- it stimulates foresight in that it enables the *Client* to make rational decisions about changes to the work with reasonable certainty of its cost and time implications
- at the same time, it puts a risk on the *Contractor* that motivates it to manage the new situation efficiently.

An important by-product of the procedure included in the ECC for compensation events is that few, if any, issues relating to the valuation of work or extensions of time are left to be settled after the event.

> The compensation event procedure is a quick procedure designed to value change during the period of the contract and not after Completion.

1.3. What is a compensation event?

Compensation events are events that are at the *Client's* risk under the Contract and that entitle the *Contractor* to an assessment of the effect that the event has on the Prices, the Completion Date and Key Dates. Risks that are not specifically identified as being the *Client's* are at the *Contractor's* liability (clause 81.1). The *Contractor* should therefore be aware of the matters that are likely to arise and will be at its risk under the contract. These items should be considered in the carrying out of its risk assessment prior to the *starting date*, and these matters should be listed in Contract Data part two to be included in the Early Warning Register.

The assessment of a compensation event is always of its effect on the Prices, the Completion Date and Key Dates. In other words, there are not separate clauses for events that result in an 'extension of time' and for events that result in changes to the Prices. For all events, the effect on both the Prices and the programme are always considered together. This does not mean that every event will always have an effect on both time and the Prices, but this effect has to be assessed in order to reach such a conclusion. In the case of some events, the assessment may be reduced payments to the *Contractor*.

> Compensation events consider the effect on both time and money.

A compensation event is a different name for various terms that do not apply in the ECC, and it should not be referred to as

- a variation
- an extension of time
- loss and expense
- delay and disruption
- a claim.

1.4. Where to find a list of compensation events

Compensation events can be found in three places:

- clause 60.1 – which lists 21 compensation events
- main Options B and D – which include additional compensation events in clauses 60.4, 60.5 and 60.6 relating to the use of *bills of quantities*
- secondary Option clauses X2.1, X10.5, X12.3(6), X12.3(7), X14.2 and X15.2, and Y(UK)2 clause Y2.5.

A compensation event is much more than Scope changes. A change in Scope is covered by only one of the 21-plus compensation events that are listed.

1.4.1 Core compensation events

A discussion of each of compensation events listed under Clause 60.1 of the ECC is given below.

Compensation event 60.1(1)

'The *Project Manager* gives an instruction changing the Scope except

- a change made in order to accept a Defect or
- a change to the Scope provided by the *Contractor* for its design which is made
  - at the *Contractor's* request or
  - in order to comply with the Scope provided by the *Client*.'

This clause is simply the result of an instruction given by the *Project Manager* varying the *works*, for example deletion or addition of work, a change to the specifications, issue of a revised drawing or clarification of a verbal instruction. Any instruction to change the *works* as a result of clause 17.1 (ambiguities and inconsistencies) or 17.2 (illegal and impossible requirements) will also fall under this compensation event.

The bullet points in clause 60.1(1) detail the three exceptions to a change to the Scope being a compensation event:

- The *Project Manager* may agree, for reasons of efficacy, to accept a Defect created by the *Contractor* (clause 45; discussed further in Chapter 3 of Book Three). If so, then they would instruct a change to the Scope after acceptance of the *Contractor's* quotation under clause 45.2. The resulting change to the Scope to ensure that the Scope reflects the *works* as built (including the Defect) is not regarded as a compensation event.
- Where the *Contractor* has designed the *works*, and/or has included the Scope as part of its proposal, there will be two parts to the Scope that form part of the contract: the Scope provided by the *Client* and the Scope provided by the *Contractor*. Although the Scope by the *Contractor* forms part of the overall Scope, the *Contractor* retains ownership of its part, and any changes to it made at the request of the *Contractor* or to ensure that it complies with the Scope provided by the *Client* are not construed as changes to the Scope provided by the *Client*. To avoid confusion, a *Project Manager* instructing a change to the *Contractor's* Scope should make it clear in the instruction that the change is not to the *Client's* Scope, but is a change to the *Contractor's* Scope either made at the *Contractor's* request or to comply with the Scope by the *Client*.
- Where an instruction is given by the *Project Manager* for the *Contractor's* design to comply with the *Client's* Scope. In this respect, it means the *Client's* Scope takes priority over the *Contractor's* design.

> The *Client's* Scope takes priority over the *Contractor's* design.

Compensation event 60.1(2)
'The *Client* does not allow access to and use of each part of the Site by the later of its *access date* and the date for access shown on the Accepted Programme.'

The *Client* will have included *access dates* in Contract Data part one; that is, dates by which the *Client* intends to give the *Contractor* access to the Site or parts of it. The *Contractor* will have included in their programme submitted for acceptance the date by which they require access to the Site or parts of it. These latter dates included by the *Contractor* in their programme may be later than the *access dates* proposed by the *Client*. Clause 33.1 clearly states the obligations of the *Client* in giving access to the Site to the *Contractor*; that is, to give access by the later of the *access date* stated in Contract Data part one or the date for access given on the Accepted Programme. This is discussed further in Chapter 2 of Book Three. If the *Client* fails in this obligation, it is a compensation event.

Compensation event 60.1(3)
'The *Client* does not provide something which it is to provide by the date shown on the Accepted Programme.'

The wording is to make it clear that where a *Contractor* requires the *Client* to provide something that this is shown on the Accepted Programme and not lost in supporting information.

The Scope should state clearly details of anything, such as Plant and Materials or facilities, that the *Client* is to provide and any restrictions on when it is to be provided. Clause 31.2 requires the *Contractor* to include this information in their Accepted Programme. If the *Client* fails to provide this information by the relevant date, the *Contractor* is entitled to notify the event as a compensation event. Note that this compensation event depends on the *Contractor* having entered the dates in their programme.

Compensation event 60.1(4)
'The *Project Manager* gives an instruction to stop or not to start any work or to change a Key Date.'

Clause 34.1 gives the *Project Manager* the authority to instruct the *Contractor* to stop or not to start work. One of the many reasons that the *Project Manager* may give such an instruction is for reasons of safety. Such an instruction is a compensation event. Some clients change this clause using a secondary Option Z clause to add that where the instruction relates to health and safety matters or is in relation to a *Contractor* default, the instruction is not a compensation event. Clause 14.3 gives the *Project Manager* the authority to instruct the *Contractor* to change a Key Date.

Compensation event 60.1(5)
'The *Client* or Others

- do not work within the times shown on the Accepted Programme,
- do not work within the conditions stated in the Scope or
- carry out work on the Site that is not stated in the Scope.'

The Scope should state clearly details of the order and timing of work to be done by the *Client* and Others. Clause 31.2 requires the *Contractor* to include this information in their Accepted Programme. If the *Client* or Others work outside these parameters, it is a compensation event. Note that this compensation event depends on the *Contractor* having entered the dates in their programme.

Compensation event 60.1(6)
'The *Project Manager* or the *Supervisor* does not reply to a communication from the *Contractor* within the period required by the contract.'

Certain clauses within the ECC give various periods for reply by the *Project Manager* and *Supervisor*. A default *period for reply* is given in Contract Data part one, and the obligation to reply within the relevant period is given in clause 13.3. Where communication is not made within the timescales given, the *Contractor* may notify a compensation event. Note that any time period may be extended by agreement between the *Project Manager* and the *Contractor*.

Compensation event 60.1(7)
'The *Project Manager* gives an instruction for dealing with an object of value or of historical or other interest found within the Site.'

Clause 73.1 states the procedure for dealing with such items. Any instruction for dealing with an object of value or of historical or other interest found within the Site would be additional work for the *Contractor* and therefore a compensation event.

Compensation event 60.1(8)
'The *Project Manager* or the *Supervisor* changes a decision which either has previously communicated to the *Contractor*.'

Both the *Project Manager* and the *Supervisor* are able to change decisions made under the authority given to them under the ECC. Any such changed decision is likely to result in extra work for the *Contractor*, and would therefore be a compensation event.

Compensation event 60.1(9)
'The *Project Manager* withholds an acceptance (other than acceptance of a quotation for acceleration or for not correcting a Defect) for a reason not stated in the contract.'

There are various clauses in the ECC that state reasons why the *Project Manager* is entitled not to accept a submission or proposal from the *Contractor*. Examples are clauses 24.1 (people) and 31.3 (the programme). Where the *Project Manager* does not accept a submission from the *Contractor* and the reason that they state is not one of the reasons in the contract, then the *Contractor* may notify a compensation event. If the withheld acceptance is for a quotation for acceleration (clause 36) or for acceptance of a Defect (clause 45), then the non-acceptance is not a compensation event. This is because both the quotation for acceleration and the quotation for accepting a Defect are voluntary.

Compensation event 60.1(10)
'The *Supervisor* instructs the *Contractor* to search for a Defect and no Defect is found unless the search is needed only because the *Contractor* gave insufficient notice of doing work obstructing a required test or inspection.'

Clause 43.1 allows until the *defects date* for the *Supervisor* to instruct the *Contractor* to search. Since a search that does not reveal a Defect would have been unfair to the *Contractor*, a compensation event allows them to notify the time and cost that the unnecessary search has resulted in. If, however, the search was required because the *Contractor* did not give sufficient notice for the test or inspection (clause 41.3), then any search is not a compensation event, whether or not a Defect is found.

Compensation event 60.1(11)
'A test or inspection done by the *Supervisor* causes unnecessary delay.'

The Scope should state clearly those tests or inspections that are to be carried out by the *Supervisor* and the *Contractor*, whether witnessed by the *Supervisor* or not. Clause 41.5 requires the *Supervisor* to carry out their tests and inspections without causing unnecessary delay. Although the word 'unnecessary' may be a little vague, the *Contractor* could use their programme for the commencement date of following activities to provide evidence of the delay.

Compensation event 60.1(12)
'The *Contractor* encounters physical conditions which

- are within the Site

- are not weather conditions and
- an experienced contractor would have judged at the Contract Date to have such a small chance of occurring that it would have been unreasonable to have allowed for them.'

Only the difference between the physical conditions encountered and those for which it would have been reasonable to have allowed is taken into account in assessing a compensation event.

The wording clarifies that the *Contractor's* entitlement is limited to the event's effect over and above that which 'would have been reasonable to have allowed'.

This compensation event means that the *Client* takes the risk for physical conditions. Note that 'physical conditions' includes more than just ground conditions. A statement in the instructions to tenderers that

> 'the tenderer shall make whatever arrangements are necessary to become fully informed regarding all existing and expected conditions and matters that might in any way affect the cost of the performance of the *works* and claims for additional reimbursement on the grounds of lack of knowledge or failure to fully investigate the foregoing conditions shall not relieve the tenderer from the responsibility for estimating properly the difficulty or cost of successfully performing any work'

does not relieve the *Client* of their responsibilities.

As with the integration of the early warning clause into a compensation event (clause 61.5), clause 60.1(12) refers to an 'experienced' contractor.

Clause 60.1(12) is read with clauses 60.2 and 60.3. Clause 60.2 describes the aspects of the physical conditions: whether they had such a small chance of occurring that it would have been unreasonable to have allowed for them (note that the compensation event refers to physical conditions and not simply ground conditions, where 'physical conditions' has a much wider connotation), and that the *Contractor* is assumed to have taken into account when judging (at the Contract Date). This means it is in the *Client's* interests to provide as much information to the *Contractor* as possible, in order to discharge their own duties.

Clause 60.3 states the *contra proferentem* rule regarding inconsistencies and ambiguities in the Site Information, for which the *Client* is responsible.

This compensation event is the standard ground conditions variation that has been with the construction industry for many years. It is important when preparing the tender documentation under the ECC that the following points are borne in mind:

- The more information concerning ground conditions that can be provided, the greater the certainty with which appropriate allowances can be made by the tenderers.
- It is important that the information provided is both correct and relevant to the risks faced.
- It may be useful for the *Client* to use a specialist to provide interpretation of factual data to ensure that tenders are on a common basis.
- The *Client* may use the Scope or secondary Option Z whereby they can define in the contract the limit between the risks carried by the *Client* and the *Contractor*; that is, to indicate what should be allowed for in the Prices. An example of such limits would be for the *Client* to state the limits for groundwater levels.

If a *Client* chooses to delete clause 60.1(12), using Option Z, then **all** the risks for physical conditions are taken by the *Contractor*, not only those that they have misjudged, given the information provided by the *Client*. Perhaps the *Client* should consider why the *Contractor* should take the risk of physical conditions if the *Client* is not prepared to, even though the *Client* is more likely to have the information regarding the physical conditions.

> Deleting compensation events 60.1(12) and 60.1(13) means more than simply deleting those clauses. It is also worth considering whether the *Contractor* can manage those risks better than the *Client*.

Compensation event 60.1(13)
'A *weather measurement* is recorded

- within a calendar month
- before the Completion Date for the whole of the *works* and
- at the place stated in the Contract Data

the value of which, by comparison with the *weather data*, is shown to occur on average less frequently than once in 10 years.

Only the difference between the *weather measurement* and the weather which the *weather data* show to occur on average less frequently than once in 10 years is taken into account in assessing a compensation event.'

The wording clarifies that the *Contractor's* entitlement is limited to the difference between the *weather data* and the *weather measurement*.

The ECC does not refer to 'inclement' weather, or 'exceptionally adverse weather conditions', but rather weather that occurs on average less frequently than once in 10 years, using a defined set of records, such as those available from the Met Office (see later in this section) or other independent body. This is a more objective and measurable approach than other standard forms of contract.

The purpose is to make available for each contract *weather data*, compiled by an independent authority (Contract Data part one requires the insertion of who is to supply the *weather measurements*) and agreed by both Parties beforehand, establishing the levels of selected relevant weather conditions for the Site for each calendar month that has had a period of return of more than 10 years. If weather conditions more adverse than these levels occur, it is a compensation event. Weather which the *weather data* show is likely to occur less frequently than once within a 10-year period is the *Contractor's* risk in relation to both cost and time.

The time of occurrence of all compensation events is when the action or lack of action describing the event takes place. In the case of weather, it is the day when weather conditions are recorded as having occurred within a calendar month and are on average more frequent than once in 10 years. The test is the comparison of the *weather measurements* with the *weather data*. The compensation event can then be notified under clause 61.3, and its effect can be assessed at the end of the month when the extent of the weather exceeding the 10-year return *weather data* is known. The process starts again at the beginning of each month.

This compensation event is concerned with weather occurring only at the place stated in the Contract Data. If weather occurring at some distance from the Site could produce some risk such as flooding on the Site, the allocation of risk should be dealt with by special compensation events.

It should be noted that the ECC awards both time and money to the *Contractor* who successfully proves a weather compensation event. Traditional contracts tend to award an extension of time but no money, and, for this reason, some *Clients* are unhappy at having to pay for an event that is not within their control in the same way that the other events are within their control. Many *Clients* have deleted this clause using Option Z. Perhaps these *Clients* are not aware of how onerous the weather compensation event actually is, and how much risk the *Contractor* is actually adopting already. This point is worth emphasising, since it is a common misconception that the ECC weather statement is less onerous than in traditional contracts.

The criterion is weather that **occurs on average less frequently than once in 10 years**. Let us say that there has been a large amount of rainfall in May and the *Contractor* wishes to notify a compensation event. They should first, having measured the rainfall at the place stated in the Contract Data (hopefully on or near the Site), average the rainfall occurring in that May. They should then access records that give the rainfall in every May for the period of return, which should be for a period of more than 10 years. (The records from the Met Office tend to give a 10-year average, making comparisons much easier.) The average rainfall for each May is then compared. If the average rainfall in the month of May during which the *Contractor* was Providing the Works was on average over the period of return greater than the one in 10-year average for May, then the compensation event may be notified. If the relevant month of May was the same as the highest May cumulative rainfall, then it does not fit the criterion, since it would then be equal to, not less than, once in 10 years. The scenarios below are based on the period of return records being available since 1983.

Through these examples it can be seen that the *Contractor* is required to measure the *weather measurements* such as rainfall and compare them with the *weather data* from the *weather data* available over the period of return. Only if the *weather measurement* by comparison to the *weather data* is shown to occur on average less frequently than once in 10 years does it qualify as a compensation event. Note that the examples above illustrate how the once in 10-year average would work.

The Met Office (www.met-office.gov.uk) offers subscription services for planning averages and monthly updates for the NEC:

> 'Location Based Reports, available as Location Based Monthly Planning Averages and Location Based Monthly Downtime Summaries, both provide weather information, from over 3,600 locations, that accurately reflects onsite conditions ... To create them, we have combined our historic gridded database of long-term average values with our database of present observations used to drive forecasting models.
>
> These new reports also include up to 16 weather parameters with wind as standard so they can be used for a variety of building contracts including New Engineering Contract (NEC) clause 60.1 (13).' (Met Office website, 2017)

It makes sense that the *Client* takes the risk for elements that are outwith the *Contractor's* control. If the *Contractor* is to take the risk, they are likely to factor this risk into the contract, and the *Client* is unlikely to know whether they are receiving value for money. The *Contractor* is likely to be conservative in their risk estimate and the true price of the project could be difficult to assess. The assessment of weather applies to the 'extra' weather and not to the one in 10-year weather that is at the *Contractor's* risk.

Compensation event 60.1(14)
'An event which is a *Client's* liability stated in these *conditions of contract*.'

The wording reflects the fact that this clause does not just refer to clause 80.1, which lists the *Client's* liabilities, but also includes any additional *Client's* liabilities stated in Contract Data part one.

Clause 80.1 lists the *Client's* liabilities. Additional *Client's* liabilities, if any exist, are stated in Contract Data part one. Those *Clients* providing design to the *Contractor* should note that a *Client's* liabilities event is claims, proceedings, compensation, and claims that are due to a fault of the *Client* or a fault in their design.

Weather data

Scenario 1

The average rainfall (in mm) for the month of May in the relevant year (in this case 2017) is 144. The period of return is based on the available *weather data* for the 30-year period 1988–2017 as follows:

2017	2016	2015	2014	2013	2012	2011	2010	2009	2008
144	125	114	130	137	141	140	120	114	130
2007	2006	2005	2004	2003	2002	2001	2000	1999	1998
137	141	127	133	140	143	130	126	124	132
1997	1996	1995	1994	1993	1992	1991	1990	1989	1988
135	137	140	141	142	120	118	121	121	112

The average rainfall in May 2017 at 144 is the highest based on the period of return weather records being available for the 30-year period since 1988 and therefore fits the criterion of occurring on average less frequently than once in 10 years. The rainfall may be notified as a compensation event under clause 60.1(13).

Scenario 2

The average rainfall for the month of May in the relevant year (in this case 2017) is 142. The period of return is based on the available *weather data* for the 30-year period 1988–2017 as follows:

2017	2016	2015	2014	2013	2012	2011	2010	2009	2008
142	125	114	130	137	140	140	124	113	143
2007	2006	2005	2004	20003	2002	2001	2000	1999	1998
143	140	127	133	**144**	140	132	122	123	133
1997	1996	1995	1994	1993	1992	1991	1990	1989	1988
132	137	141	**145**	140	121	119	121	121	112

The average rainfall in May 2017 at 142 is the fourth highest based on the period of return weather records being available since 1988 behind 1994 (145), 2003 (144) and 2007 (143) based on the period of return weather records being available since 1988. A simplistic approach is to estimate the once in 10-year value for the available weather records in this scenario for 30 years and to take the third highest to estimate the one in 10 years average in this case (2007: 143).

The average rainfall in May 2017 is 142, which is the fourth highest average over the period of return and therefore does not fit the criterion of occurring on average less frequently than once in 10 years. The rainfall may not be notified as a compensation event under clause 60.1(13).

Scenario 3

The average rainfall for the month of May in the relevant year (in this case 2012) is 143. The period of return is based on the available *weather data* for the 30-year period 1983–2012 as follows:

2017	2016	2015	2014	2013	2012	2011	2010	2009	2008
143	125	114	130	137	141	**143**	124	112	143
2007	2006	2005	2004	2003	2002	2001	2000	1999	1998
135	140	127	133	140	143	**143**	122	123	133
1997	1996	1995	1994	1993	1992	1991	1990	1989	1988
132	137	141	**143**	142	121	119	120	119	122

The average rainfall in May 2017 is 143, but this highest value has occurred previously during the period of return between 1988 and 2017 in 2011, 2001 and 1994. The rainfall in May 2017 at 143 does not fit the criterion of occurring on average less frequently than once in 10 years, as on average over the 30-year period of return it has occurred four times. The rainfall may not be notified as a compensation event under clause 60.1(13).

Compensation event 60.1(15)

'The *Project Manager* certifies take over of a part of the *works* before both Completion and the Completion Date.'

The *Client* may use a part of the *works* before Completion and, unless the use is for reasons stated in Clause 35.2, they take over that part. If take over occurs before Completion **and** the Completion Date, it is a compensation event.

It is important to note that the Scope should state the reasons, if any exist, as to why the *Client* may require to use part of the *works* before Completion. For example, the *Client* could require access across parts of the *works* for their own requirements. Alternatively, the *Contractor* may request the *Client* to use part of the *works* to suit their method of working. Under clause 35.2, take over would **not** occur in either of these instances, and therefore there would be no compensation event. The *Client* may state in Contract Data part one that they are unwilling to take over the *works* before the Completion Date (to cover instances where the *Contractor* completes early and expects the *Client* to take over early).

Compensation event 60.1(16)

'The *Client* does not provide materials, facilities and samples for tests and inspections as stated in the Scope.'

Clause 41.2 requires the *Client* to provide materials, facilities and samples for tests and inspections as stated in the Scope. This compensation event relies on the Scope stating the things that the *Client* is to provide. If the *Client* does not provide the things they are required to provide, then the *Contractor* is entitled to notify a compensation event.

Compensation event 60.1(17)

'The *Project Manager* notifies the *Contractor* of a correction to an assumption which the *Project Manager* made about a compensation event.'

The wording reflects that the compensation event only relates to assumptions made by the *Project Manager* and not to those made by the *Contractor*.

Clause 61.6 allows the *Project Manager* to state assumptions to be used to facilitate the assessment of a compensation event. If they later notify the *Contractor* of corrections to these assumptions, the notification is a separate compensation event.

Compensation event 60.1(18)

'A breach of contract by the *Client* which is not one of the other compensation events in this contract.'

This is an 'umbrella' clause to include breaches of contract by the *Client* within the compensation event procedure.

Compensation event 60.1(19)

'An event which

- stops the *Contractor* completing the whole of the *works* or
- stops the *Contractor* completing the whole of the *works* by the date for planned Completion shown on the Accepted Programme

and which

- neither Party could prevent
- an experienced contractor would have judged at the Contract Date to have such a small chance of occurring that it would have been unreasonable to have allowed for it and
- is not one of the other compensation events stated in this contract.'

This compensation event deals with events where the chances of it happening are so remote as to be unreasonable to have provided for it in the contract.

No attempt has been made as with other contracts to define what such an event is (war, act of God, etc.). A legal definition of *force majeure* is as follows:

> '**Force majeure** [French: Irresistible compulsion or coercion. The phrase is used particularly in commercial contracts to describe events possibly affecting the contract and that are completely outside the parties' control. Such events are normally listed in full to ensure their enforceability; they may include acts of God, fires, failure of suppliers or subcontractors to supply the supplier under the agreement, and strikes and other labour disputes that interfere with the supplier's performance of an agreement. An express clause would normally excuse both delay and a total failure to perform the agreement.' (*Oxford Dictionary of Law*, 7th edn, 2014, Oxford University Press)

This clause also makes it a positive obligation on the *Contractor* to notify such events.

An example of a situation where this compensation event could be used is a foot-and-mouth epidemic. The epidemic that occurred in 2001 in the UK had a severe effect on pipe laying and other projects in areas affected by the disease.

Compensation event 60.1(20)
'The *Project Manager* notifies the *Contractor* that a quotation for a proposed instruction is not accepted.'

The *Contractor* is entitled to compensation for a proposed instruction not being accepted. This will encourage the *Project Manager* to consider carefully issuing proposed instructions, which, in the past, may have resulted in the *Contractor* incurring costs for which there was no contractual means of compensation unless a quotation for a proposed instruction was accepted.

Compensation event 60.1(21)
'Additional compensation events stated in Contract Data part one.'

This provides an opportunity for the *Client* to identify additional compensation events.

Why include additional compensation events?
- In order to get better value for money from the *Contractor* by defining what could be a grey area (e.g. physical conditions) or simply taking on board a risk. This could be defined up-front or perhaps as a result of tender negotiations.

Examples of additional compensation events:
- Discovery of archaeological finds
- Discovery of asbestos
- Birds nesting
- Unexploded bombs
- Discovery of the great crested newt
- Suicide attempts off a bridge
- Late flights
- Late-running trains
- Planning approval.

What is the consequence?
- If they occur, then they would become a compensation event (clause 60.1(21)).
- The *Client* also indemnifies the *Contractor* (clause 83.1).
- The *Contractor* does not have to insure against a *Client's* liabilities (clauses 80.1 and 83.1).

1.4.1.1. Summary of clauses

Table 1.1 gives an at-a-glance summary of the clauses referred to in the compensation events.

Table 1.1 Summary of compensation events

Compensation clause	Brief description	Relevant clause
60.1(1)	Change to the Scope	14.3, 27.3 and 45
60.1(2)	Access and use of the Site	33.1
60.1(3)	*Client* providing something	31.2
60.1(4)	Stop or not start any work	34.1
60.1(5)	*Client* and Others working times and conditions	31.2
60.1(6)	Replying to communications	13.3
60.1(7)	Object of value	73.1
60.1(8)	Changing decisions	No specific clause
60.1(9)	Withholding acceptance	13.8 and, for example 13.4, 24.1 and 31.3
60.1(10)	Instructions to search	43.1
60.1(11)	Test or inspection causing delay	41.5
60.1(12)	Physical conditions	See also clauses 60.2 and 60.3
60.1(13)	Weather	No specific clause/Contract Data part one
60.1(14)	*Client's* liabilities	80.1
60.1(15)	Take over	35.2
60.1(16)	*Client* provides materials, facilities and samples	41.2
60.1(17)	Correction to an assumption	61.6
60.1(18)	Breach of contract	No specific clause
60.1(19)	Unforeseen events	19.1
60.1(20)	Quotation for a proposed instruction is not accepted	65
60.1(21)	Additional compensation events	Contract Data part one

1.4.2 Main Options B and D only

Since Options B and D are based on Bills of Quantities that are at the *Client's* risk, there are three additional compensation events applicable to these main Options only under the contract, as shown in Table 1.1.

Compensation event 60.4

'A difference between the final total quantity of work done and the quantity stated for an item in the Bill of Quantities is a compensation event if

- the difference does not result from a change to the Scope
- the difference causes the Defined Cost per unit of quantity to change and
- the rate in the Bill of Quantities for the item multiplied by the final total quantity of work done is more than 0.5% of the total of the Prices at the Contract Date.

If the Defined Cost per unit of quantity is reduced, the affected rate is reduced.'

The ECC sets an objective test of when a change in quantity leads to a change in rate in the Bill of Quantities.

This clause only applies to changes in quantities that do not result from changes to the Scope. A change to the Scope is always a compensation event, subject to the exceptions in clause 60.1(1), regardless of the effect on quantities.

A change in quantity is not, in itself, a compensation event. A compensation event is triggered only by the changed quantity satisfying the three tests stated in the clause.

So, by way of an example, if in a Bill of Quantities for earthworks the final total quantity of work done compared with the item stated in the bill is more than 0.5% of the total of the Prices, it does not arise from a change in the Scope and the change causes the affected rate to change, then this difference would be assessed as a compensation event.

Scenario A. The final total quantity of work done is

Reference	Item description	Unit	Quantity	Rate	Amount: £
6.01	**Series 600: EARTHWORKS** **Excavation** Excavation of unacceptable material Class U1A in cutting and other excavation 0 to 3 m in depth	m^3	10 000	10.00	100 000.00

and the quantity stated for an item in the Bill of Quantities is

Reference	Item description	Unit	Quantity	Rate	Amount: £
6.01	**Series 600: EARTHWORKS** **Excavation** Excavation of unacceptable material Class U1A in cutting and other excavation 0 to 3 m in depth	m^3	5000	10.00	50 000.00

and the total of the Prices at the Contract Date was £1 000 000.00.

In this scenario the change is above the trigger value, as 0.5% of the total of the Prices at the Contract Date is (£1 000 000) £5000.00, and the total value change is + £5000.00.

If we assume that the change in quantity enables the *Contractor* to work more economically and the Defined Cost per unit to be reduced, then the affected rate reduces from £10.00 per m^3 to £9.00 per m^3.

Scenario B. The final total quantity of work done is

Reference	Item description	Unit	Quantity	Rate	Amount: £
6.01	**Series 600: EARTHWORKS** **Excavation** Excavation of unacceptable material Class U1A in cutting and other excavation 0 to 3 m in depth	m^3	5000	10.00	50 000.00

and the quantity stated for an item in the Bill of Quantities is

Reference	Item description	Unit	Quantity	Rate	Amount: £
6.01	**Series 600: EARTHWORKS** **Excavation** Excavation of unacceptable material Class U1A in cutting and other excavation 0 to 3 m in depth	m^3	10 000	10.00	100 000.00

and the total of the Prices at the Contract Date was £1 000 000.00.

In this scenario the change is above the trigger value, as 0.5% of the total of the Prices at the Contract Date is (£1 000 000) £5000.00, and the total value change is −£5000.00.

If we assume that the change in quantity means that the *Contractor* is not able to work as economically and the Defined Cost per unit will rise, this causes the affected rate to increase to £12.00 per m^3 from £10.00 per m^3.

Compensation event 60.5
'A difference between the final total quantity of work done and the quantity for an item stated in the Bill of Quantities which delays Completion or the meeting of the Condition stated for a Key Date is a compensation event.'

A difference between the original and final quantities in a Bill of Quantities is not, in itself, a compensation event. The amount due to the *Contractor* includes the Price for Work Done to Date, which is based on the actual quantities of work done. However, any difference of quantities, which causes Completion to be delayed or delays a Key Date, is a compensation event.

Compensation event 60.6
'The *Project Manager* gives an instruction to correct a mistake in the Bill of Quantities which is

- a departure from the rules for item descriptions or division of work into items in the *method of measurement* or
- due to an ambiguity or inconsistency.

Each such correction is a compensation event which may lead to reduced Prices.'

Since the Bill of Quantities is not the Scope (clause 56.1), any mistakes in the Bill of Quantities arising because the bill does not comply with the *method of measurement* or because of ambiguities or inconsistencies are treated separately (clause 17.1). This may occur when an item has been omitted from the bill or an item in the bill should be deleted or amended to comply with the *method of measurement*. This is one of the compensation events that may result in a reduction of the Prices.

1.4.3 Secondary Options X2, X10, X12, X14, X15 and Y(UK)2

The inclusion of the following secondary Options gives rise to additional compensation events.

Option X2 – changes in the law
If, after the Contract Date, a change in the *law of this contract* occurs, it is a compensation event.

Option X10 – building information management
If the Information Execution Plan is altered by a compensation event, the *Contractor* includes alterations to the Information Execution Plan in the quotation for the compensation event (clause X10.5).

Option X12 – changes in the law
A change to the Partnering Information (clause X12.3(6)) or to the *Contractor*s programme (clause X12.3(7)) is a compensation event, which may lead to reduced Prices.

Option X14 – advanced payment to the *Contractor*
If there is a delay in the *Client's* making the advanced payment under this secondary Option, a compensation event occurs (clause X14.2).

Option X15 – limitation of the *Contractor's* liability for their design to reasonable skill and care
If the *Contractor* corrects a Defect for which they are not liable under the contract, it is a compensation event (clause X15.2).

Option Y(UK)2 – Part II of the Housing Grants, Construction and Regeneration Act 1996
Suspension of performance is a compensation event if the *Contractor* exercises their right to suspend performance under the Act (clause Y2.5).

1.5. Roles of the *Project Manager* and *Contractor*

The roles of the *Project Manager* and *Contractor* in the compensation event procedure are defined in the ECC in terms of the actions each is to take. These actions are all described in section 6 of the core clauses – 'Compensation events' (Appendix 1, Sections A and B in this book list those actions required by the *Project Manager* and *Contractor* with regard to compensation events).

The list of compensation events contained within clause 60.1 divide generally into two categories:

- those for which the *Project Manager* will usually volunteer a decision that a compensation event has occurred; that is, those generally in respect of instructions issued by the *Project Manager* or *Supervisor* (see Section 1.7.2 below)
- those that, due to their subjectivity or because the event could be construed as some shortcoming of the *Project Manager*, *Supervisor* or *Client*, are more likely to be left to the *Contractor* to notify the *Project Manager* that they consider a compensation event has occurred (see section 1.7.3 below).

Appendix 1, Sections D and E in this book shows the different procedures to be followed in each case. It will be seen that the procedures are identical once the *Project Manager*, for events in the second of the above categories, notifies the *Contractor* that they believe a compensation event has occurred.

The flow charts in Appendix 1 show the 'trouble-free' situations where the procedure operates smoothly. In practice, however, complicating factors can arise to disrupt the smooth process envisaged, but which in fairness to the ECC have been predicted and provided for in the contract. Appendix 1, Section F, presents a table of the more common complicating factors that can arise and some possible consequences.

1.6. Administering compensation events

The assessment of a compensation event is always of its effect on both cost (the Prices) and the programme (the Completion Date and Key Dates).

1.6.1 Changes to the Completion Date

The *Contractor* includes alterations to the Accepted Programme as part of their quotation for the compensation event where the programme has altered in any way. A change to the programme includes not only a change to the Completion Date or Key Dates but changes to the resources, the statement of how the *Contractor* plans to do the work or sequencing of the programme. In any of these cases, alterations to the Accepted Programme are required to be submitted with the quotation.

> Compensation events are a package of time and money; therefore, the programme is a part of a quotation for a compensation event.

The programme is an important part of a compensation event quotation. The *Contractor* should ensure that all changes are noted in the programme, including consequential changes that result from a compensation event.

1.6.2 Changes to the Prices

Clause 63.1 states that the changes to the Prices are assessed as the effect of the compensation event upon the Defined Cost of the work already done, the forecast Defined Cost of the work not yet done and the resulting Fee.

The important principle here is that there is absolutely no reference to the Prices included in the *activity schedule* (clause A 11.2(21) – Activity Schedule is a defined term): the Prices in main Options C and D are used to determine the target price only. Instead, assessment of the financial effects of a compensation event is based on their effect on Defined Cost plus the Fee.

Defined Cost is defined in all the main Options. The Fee is the amount calculated by applying the *fee percentage* stated in Contract Data part two to the amount of Defined Cost, and is intended to cover such items as the *Contractor's* offsite overheads, profit and any other cost components not expressly included in the Schedule of Cost Components (SCC or Short SCC) (see clause 52.1).

What this all means is that for every compensation event a 'mini-lump sum' price is, wherever possible, estimated in advance, and is based on its forecast effect on

- the actual Defined Cost or
- the actual Defined Cost of the work already done, if the assessment is made after the work because the subject of the compensation event has been completed.

No compensation event for which a quotation is required is due to the fault of the *Contractor* or relates to a matter that is at their risk under the contract. It is therefore appropriate to reimburse the *Contractor* their forecast additional costs. Clause 63.1 identifies that the demarcation between the actual Defined Cost of the work already done and the forecast Defined Cost of work yet to be done is the date that the *Project Manager* instructed or should have instructed the work.

1.6.3 Procedure for change
1.6.3.1 Forecasts

It is the intention of the ECC that the majority of quotations for compensation events are based on forecasts of their financial effects (provided by the timescales included in the contract), since this accords with the objective designed to provide the *Client* with reasonable certainty of the cost and time implications of changes to the work and places a risk on the *Contractor*, which motivates the latter to manage the new situation efficiently.

Where the effects of a change are too uncertain to be forecast reasonably by the *Contractor*, the *Project Manager* states assumptions about the event on which the *Contractor* bases their forecast of Defined Cost. This precludes the use of large contingent sums in the *Contractor's* quotations. The *Project Manager's* assumptions provide the only mechanism for revisiting the compensation event quotation after implementation.

1.6.3.2 Revisiting compensation events

The quotation provided by the *Contractor* is their **only** chance of including **all** costs (consequential or otherwise) resulting from a particular event. Once a compensation event has been implemented, only the *Project Manager's* assumptions that turn out to be incorrect allow the revisiting of a quotation (clauses 61.6 and 60.1(17)). Any assumptions made by the *Contractor*, if later proved to be incorrect, do not allow such reassessment. This must be made abundantly clear to *Contractors*.

1.6.3.3 Times stated in the procedure

In practice, many compensation events can occur simultaneously, and some compensation events may involve significant restructuring of the price document (and programme). In recognition of this the ECC allows the relaxation of the time periods by agreement between the *Project Manager* and the *Contractor* (clause 62.5). Such relaxation, however, should be the exception to the rule and not used as a cover-up for ineffective or poor contract administration.

It may be the case that the *Project Manager* has been given a specific level of financial authorisation and that they would be required to report to a management board on increases to the project value. They could therefore find it difficult to respond in the time required by the contract where a compensation event has breached the maximum level of their financial authority, whether per event or total contract value. This would apply particularly to the 2-week reply to quotations (clause 62.3).

Where the *Client* wishes to initiate specific procedures covering this scenario, amendments to the *conditions of contract* can be made using Option Z. For one-off cases, the *Project Manager* should rely on clause 62.5 to extend the time required.

Variations in the *Contractor's* supply chain could also result in the *Contractor* needing an extension to the time period of 3 weeks to submit a quotation for a compensation event.

In general, agreement on timescales should probably be made for each compensation event as it arises, so that the right resolution is arrived at practically while still reflecting the spirit of the contract.

Managing Change

1.6.3.4 Other aspects of the procedure

The compensation event process has specific procedural requirements, which fall into three distinct stages, namely

1. notification
2. quotation
3. implementation.

> Quotations cannot be revisited unless they are based on assumptions given by the *Project Manager* and they later correct incorrect assumptions.

1.7 Notification of a compensation event

1.7.1 Proposed instruction or changed decision

Note that although this sub-heading is included under the heading of a compensation event, the instruction of such a quotation is **not** a compensation event in itself, and the *Contractor* does not put the proposed instruction into effect.

The *Project Manager* may instruct the *Contractor* to submit quotations for a proposed instruction or a proposed changed decision (clause 65.1). The *Project Manager* states in the instruction the date by which the proposed instruction may be given. The quotation is to be submitted within 3 weeks of being instructed to do so by the *Project Manager* (clause 65.2), and the quotation is assessed as a compensation event. The *Project Manager* replies to the *Contractor's* quotation by the date when the proposed instruction may be given.

The instruction is **not** related to a compensation event as such but only to a potential compensation event, and the *Project Manager* should make this very clear in their instruction. This option is available to the *Project Manager* where they may be considering a change but wish first to know what the effect of that change would be. Once they have received the no-obligation quotation, the *Project Manager* makes one of three replies:

- an instruction to submit a revised quotation including the reasons for doing so
- the issue of the instruction together with a notification of the instruction as a compensation event and acceptance of the quotation
- a notification that the quotation is not accepted.

If the *Project Manager* does not reply to the quotation within the time allowed, the quotation is not accepted.

> The *Project Manager* may instruct the *Contractor* to submit a quotation for a proposed instruction; for example, changing the floor finishes to the entrance of their new office block.

1.7.2 Notification by the *Project Manager*

Either the *Contractor* or the *Project Manager* can notify a compensation event. There are eight instances in which the *Project Manager* should identify the compensation event (clause 61.1):

- The *Project Manager* gives an instruction changing the Scope (clause 60.1(1)).
- The *Project Manager* gives an instruction to stop or not to start any work or to change a Key Date (clause 60.1(4)).
- The *Project Manager* gives an instruction for dealing with an object of value or of historical or other interest found within the Site (clause 60.1(7)).
- The *Project Manager* or the *Supervisor* changes a decision they have previously communicated to the *Contractor* (the assumption is made that the *Project Manager* is aware of the *Supervisor's* changing a decision, presumably because they have been copied in on correspondence or because the *Supervisor* has informed them of the change) (clause 60.1(8)).
- The *Supervisor* instructs the *Contractor* to search, and no Defect is found (unless the search is needed only because the *Contractor* gave insufficient notice of doing work obstructing a required test or inspection) (clause 60.1(10)).

- The *Project Manager* certifies take over of a part of the *works* before both Completion and the Completion Date (unless this is for a reason stated in the Scope or to suit the *Contractor's* way of working) (clause 60.1(15)).
- The *Project Manager* notifies a correction to an assumption about the nature of a compensation event (clause 60.1(17)).
- The *Project Manager* notifies the *Contractor* that a quotation for a proposed instruction is not accepted (clause 60.1(20).

If the *Project Manager* does not notify a compensation event, the *Contractor* may do so (clause 61.3; if the *Contractor* does not notify of a compensation event within 8 weeks of becoming aware of the event, it is not entitled to the compensation event unless the *Project Manager* should have notified the event to the *Contractor* but did not). Although the *Project Manager* may therefore rely on the *Contractor* to notify all compensation events, even those that the *Project Manager* should notify, it falls within the boundaries of mutual trust and co-operation that the *Project Manager* notifies those compensation events that they are required to notify.

> If the *Project Manager* does not notify a compensation event, the *Contractor* may do so.

1.7.3 Notification by the *Contractor*

The *Contractor* may notify a compensation event under the following circumstances (clause 61.3):

- the *Contractor* believes the event is a compensation event
- the *Project Manager* has not notified the event to the *Contractor*.

Note that both statements have to be satisfied before the *Contractor* may notify a compensation event to the *Project Manager*. The 8 weeks within which the *Contractor* should notify a compensation event may become an indefinite period if it is a compensation event that the *Project Manager* should have notified but did not (clause 61.3). This is an encouragement to the *Project Manager* to notify compensation events. It could also potentially result in the traditional 'claims' situation, where compensation events are notified so long after the actual event that it is difficult to assess its impact.

1.7.3.1 Believing the event is a compensation event

The *Contractor* would usually believe that an event is a compensation event if it or its *works* have been affected in some way. At this stage the *Contractor* does not need to confine it to the compensation events in the contract, although clearly not doing so may lead to the notification failing the five-point test of clause 61.4.

1.7.3.2 Less than 8 weeks since they became aware of the event

This is an objective test that may be evidenced by documentation. The ECC sets this time limit to force issues to the fore and ensure that they are dealt with promptly, thereby maintaining the certainty of the final outcome, which is vital to upholding a good working relationship between the parties throughout the contract.

> The *Contractor* has only 8 weeks to notify a compensation event from becoming aware of it. This rule does not apply to events that the *Project Manager* should have notified to the *Contractor* but did not.

'If the *Contractor* does not notify a compensation event within 8 weeks of becoming aware that the event has happened, the Prices, the Completion Date or a Key Date are not changed unless the event arises from the *Project Manager* or the *Supervisor* giving an instruction or notification, issuing a certificate or changing an earlier decision.'
(clause 61.3)

'If the *Project Manager* fails to reply to the *Contractor's* notification of a compensation event within the time allowed, the *Contractor* may notify the *Project Manager* of that failure. If the failure continues for a further two weeks after the *Contractor's* notification it is treated as acceptance by the *Project Manager* that the event is a compensation event and an instruction to submit quotations.' (clause 61.4)

1.7.3.3 The *Project Manager* has not notified the event

If the *Project Manager* has notified the events that they should have notified in accordance with clause 61.1, then the events that are left to the *Contractor* to identify are

- a failure by the *Client*, *Project Manager*, *Supervisor* or Others to fulfil their obligations (compensation events 2, 3, 5, 6, 11, 16, 18, 19 and 21)
- the *Project Manager* withholding an acceptance for a reason not stated in the contract (compensation event 9)
- an event that is a *Client's* liability stated in the *conditions of contract* (compensation event 14)
- a happening not caused by any Party (compensation events 12, 13 and 19)
- events confined to the main and secondary Options and any additional compensation events stated in the Contract Data.

In reality, however, the *Contractor* would be advised to notify all events that they consider to be compensation events, even those that the *Project Manager* is supposed to notify but does not.

The occurrence of a compensation event entitles the *Contractor* to an **assessment** of time and money (as opposed to immediately entitling them to time and money; the assessment might be zero). The *Contractor's* notifying an event does not necessarily mean that they will receive time and money for the event because there is still the five-point test carried out by the *Project Manager* on receiving a notification from the *Contractor* in clause 61.4.

1.7.4 The five-point test

Once the *Contractor* has notified a compensation event, the *Project Manager* assesses the event against a five-point test as follows (clause 61.4):

Question	Yes	No
■ Does the event arise from a fault of the *Contractor*?		✓
■ Has the event happened or is it expected to happen?	✓	
■ Has not been notified within the timescales set out in the *conditions of contract*	✓	
■ Does the event affect Defined Cost, Completion or meeting a Key Date?	✓	
■ Is the event one of the compensation events stated in this contract?	✓	

Note that all five parts of the test have to be passed. Note also that if a tick were to be placed in the other column, the test would fail.

Whether the event will ever happen is a matter of opinion, as is whether the Prices, Completion or Key Dates would be affected. The first and the last points are reasonably objective, however.

If the notification passes the test(s), then the *Project Manager* instructs the *Contractor* to submit quotations for the event. If the notification fails the test, then the *Project Manager* informs the *Contractor* that the Prices, the Completion Date and the Key Dates will not be changed and the compensation event procedure ends. Of course, if the *Contractor* is unhappy with the *Project Manager's* decision, they may take the matter to adjudication.

Compensation events notified by the *Project Manager* do not go through this test. This is presumably because the *Project Manager* has already decided that the event was not the *Contractor's* fault and the event has already happened, and it is one of the compensation events stated in the contract. Whether the Prices and the Completion Date and the Key Dates will be affected will be determined after the quotation has been received.

Compensation events

1.8. Quotations for a compensation event

1.8.1 Introduction

A quotation is a time and money 'package' of the *Contractor's* assessment (unless it is a *Project Manager's* assessment) of the financial and time effects of the compensation event, and should be submitted within 3 weeks (or some other such agreed period) of the *Project Manager's* instruction to do so.

1.8.2 When are quotations submitted?

There are three instances in which a *Contractor* may be instructed by the *Project Manager* to submit quotations in relation to compensation events (the *Contractor* may also be required to submit quotations for acceleration (clause 36.1) or for the acceptance of a Defect (clause 45.2), in which case the quotations are submitted within the *period for reply*):

1 The *Project Manager* instructs the *Contractor* to submit quotations for a compensation event at the same time as they notify the compensation event (clause 61.1).
2 The *Project Manager* instructs the *Contractor* to submit quotations for a compensation event once they have decided that an event notified by the *Contractor* has passed the five-point test (clause 61.4.). Note that this is in the alternative to instance 1 above; the two cannot happen for the same event.
3 The *Project Manager* instructs the *Contractor* to submit a **revised** quotation for a compensation event (clause 62.4).

In all three instances, the *Contractor* has 3 weeks within which to submit quotations (clause 62.3 for instances 1 and 2; clause 62.4 for instance 3), but this time may be extended by agreement between the *Contractor* and the *Project Manager* before the quotation is due (clause 62.5; see additional requirements in clause 62.6). If the *Contractor* does not submit their quotation and its accompanying details within the required time, the *Project Manager* will assess the compensation event themselves, a powerful disincentive for the *Contractor*.

In addition to quotations as a result of a compensation event notification, the *Project Manager* may also instruct the *Contractor* to submit quotations for a proposed instruction (or changed decision), as discussed in Section 1.7.1 above, which could also result in a compensation event.

1.8.3 Instructions for quotations

An instruction to submit a quotation could include the following:

- A notification that the *Contractor* did not give an early warning of an event that an experienced contractor could have given (clause 61.5).
- Assumptions about the event where the *Project Manager* decides that the effects of a compensation event are too uncertain to be forecast reasonably (clause 61.6).
- An instruction to submit alternative quotations based on different ways of dealing with the compensation event. The *Project Manager* must first discuss the different ways of dealing with the compensation event that are practicable. This is particularly useful where the *Project Manager* wishes to retain the Completion Date. Where a compensation event is likely to result in a delay to the Completion Date but the *Project Manager* is keen to retain the Completion Date, rather than accelerating the *Project Manager* could request alternative quotations for the compensation event, retaining the Completion Date in one of the alternatives.

1.8.4 What is included in the quotation?

Quotations for compensation events comprise (clause 62.2)

- proposed changes to the Prices
- any delay to the Completion Date
- any delay to the Key Dates

assessed by the *Contractor*.

The quotations therefore include the following:

- Details of the assessment of the changes to the Prices and the delay to the Completion Date or Key Date
- Alterations to the Accepted Programme showing the effect of the compensation event where the programme for the remaining work has been affected (note that the programme is required if the remaining work is affected, not only if the Completion Date

has changed, in other words, if a method or resource statement has changed, sequencing amended or durations affected, a revised programme is required)
- cost and time risk allowances for matters that have a significant chance of occurring and are not compensation events (clause 63.8)
- alternative quotations where instructed to do so by the *Project Manager* (clause 61.6)
- alternative quotations for other methods of dealing with the compensation event that the *Contractor* considers practicable (clause 62.1).

The *Contractor* may include alternative quotations.

> If the *Project Manager* has notified the *Contractor* in their instruction to submit quotations that the *Contractor* did not give an early warning that an experienced contractor could have given, then the *Contractor* assesses the quotation as if they had given an early warning (clause 63.7). This is the sanction on the *Contractor* for not following the early warning procedure in the contract. The event is therefore assessed as if the options that would have been available at the time that an early warning could have been given are still available, and the quotation is for the most effective and economical option.

The quotation is based on the assumption that

- the *Contractor* reacts competently and promptly to the compensation event
- the additional Defined Cost and time due to the event are reasonably incurred
- the time due to the event is reasonably incurred (clause 63.9).

1.8.4.1 Changes to the Prices

Changes to the Prices are assessed as the effect of the compensation event upon (clause 63.1)

- the actual Defined Cost of the work by the dividing date
- the forecast Defined Cost of the work not done by the dividing date
- the resulting Fee.

Defined Cost
The first thing to notice is that all payment mechanisms deal with compensation events in the same way; that is, all compensation events are assessed using Defined Cost as defined, which, for all the main Options except Option F, are the SCC (Options C, D and E) and the Short SCC (SSCC) (Options A, B). This means that for Options A and B, which use an *activity schedule* and a Bill of Quantities as the payment mechanism rather than Defined Cost, changes and other compensation events are not assessed using the *Activity Schedule* or the Bill of Quantities. Compensation events under Options A and B are assessed using the SSCC. The *Project Manager* and *Contractor* may by agreement use rates or lump sums to assess the change to the Prices.

The effect of the event
Second, the assessment is the **effect** of the compensation event upon the three items of cost identified in clause 63.1. The event is therefore the basis for all assessment, and its effect needs to be determined.

Work already done
The calculation includes the effect upon work already done and the effect upon work not yet done. The work already done may refer to work that has been completed and that is now required to be changed. It is not the bill cost (Option B) or the activity cost (Option A) that is assessed, but the Defined Cost of the work. The chances of a *Contractor* having maintained records from which this information can be extracted are possibly small, and compiling this information could be taxing. The calculation is that the Defined Cost of the work done is deducted and the forecast Defined Cost of the new work is added.

Forecast of work not yet done
The calculation of work not yet done is a forecast. The assumption is that the work has not yet been done, although, since the *Contractor* is required to carry out the work and given the time

periods of the compensation event procedure, it is possible that the work to be done may already have been completed by the time for the submission of the quotation. This would certainly make it easier for the *Contractor* when preparing their quotation. It would reduce some of their risk, and in Options C and D would keep the target cost more stable in comparison with the Price for Work Done to Date.

Where this is not the case, the *Contractor* is required to forecast the cost using the full SCC and any Subcontractor quotations. Assumptions about the event made by the *Project Manager* could assist in this forecast.

The Fee

The Fee is defined as the amount calculated by applying the *fee percentage* to the amount of Defined Cost (clause 11.2(10)). The percentage is tendered by the *Contractor*, and included in their Contract Data part two. It represents the *Contractor's* profit and the overheads that are not included elsewhere in the SCC. Once the *Contractor* has added all the components of cost in the SCC or the SSCC, and deducted Disallowed Cost (C, D, E and F) to get their total Defined Cost, they multiply this Defined Cost by the *fee percentage* and add this product to the Defined Cost.

> The Fee includes all the costs not included in the SCC or SSCC.

1.8.4.2 Delay to the Completion Date

A delay to the Completion Date is assessed as the length of time that, due to the compensation event, planned Completion is later than planned Completion as shown on the Accepted Programme (clause 63.5). In other words, the *Contractor* includes in their programme the date when they plan to complete, as well as the date they are required to complete in accordance with the contract. This planned Completion date must be earlier than the contractual Completion Date. The duration between planned Completion and the Completion Date is the terminal float (Figure 1.1), and this remains the *Contractor's* to use if, for example, inefficiencies occur that delay planned Completion up to the Completion Date.

Figure 1.1 Terminal float

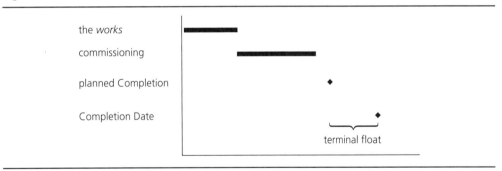

For the purposes of compensation events, therefore, the assessment of the time element of the event is based upon **planned Completion** rather than the Completion Date.

1.8.4.3 Delay to the Key Dates

Clause 63.5 also refers to the delay to a Key Date. A delay to a Key Date is assessed as the length of time that, due to the compensation event, the planned date when the Condition stated for a Key Date to be met is later than the date shown in the Accepted Programme. The *Contractor* is required to include Key Dates in their programme; that is, the dates by when the *condition* stated in Contract Data part one will be met. A compensation event quotation should include any delay to the Key Date as a result of the compensation event.

1.8.5 Acceptance of a quotation

The *Project Manager* is required to respond within 2 weeks of the *Contractor* submitting their quotation (clause 62.3), but this time may be extended by agreement between the *Contractor* and the *Project Manager* before the reply is due (clause 62.5). Their reply may be one of three options (clause 62.3):

- a notification of acceptance of the quotation
- an instruction to submit a revised quotation or
- that the *Project Manager* will be making their own assessment.

These three options are further explained as follows:

- The *Project Manager* could accept the quotation. In this case, the compensation event is implemented (clause 65.1).
- The *Project Manager* could give the *Contractor* an instruction to submit a revised quotation after explaining their reasons for doing so (clause 62.4). The *Contractor* has 3 weeks to submit the revised quotation (clause 62.4). The *Project Manager* may use this option where, for example, they consider that the *Contractor* has assessed the event incorrectly.
- The *Project Manager* could give the *Contractor* a notification that they will be making their own assessment. They could do this if, for example, they have already instructed a revised quotation but it was also unsatisfactory, or if they did not believe that a revised quotation would yield the results expected.

1.9. Assessment of quotations

The assessment of the compensation event is assumed to be carried out by the *Contractor*. The *Contractor* is required to assess the event as if they had given an early warning (where so notified), and to include time and cost risk allowances.

It is only if the *Project Manager* replies to a submitted quotation that they will be making their own assessment of a compensation event. The *Contractor* therefore always gets the first chance at assessing the event.

1.9.1 Assessment by the *Project Manager*

It is not really in either Party's interest for the *Project Manager* to assess a compensation event. The *Project Manager* will use the tools available to them, such as the Accepted Programme, which means that the *Contractor* is incentivised to keep its programme accurate and up to date. Since the reasons for the *Project Manager* making their own assessment all originate in some failure of the *Contractor*, it is possible that the *Project Manager* may be a bit more stringent in their calculations than the *Contractor*. It is unlikely that the *Contractor* will be happy about this arrangement, but its recourse is through adjudication. It therefore seems sensible that the *Contractor* prevents the compensation event procedure going as far as the *Project Manager* having to do their own assessment.

1.9.1.1 Reasons for the *Project Manager* assessing a compensation event

There are four reasons why a *Project Manager* assesses (note that there is no option for the *Project Manager* to choose not to assess if the reasons exist; the statement is obligatory) a compensation event after they have notified the *Contractor* that they will be doing so (clause 64.1):

- The *Contractor* has not submitted their quotations and accompanying details within the time allowed. The *Contractor* has 3 weeks to submit quotations after being instructed to do so, or an agreed extended time period (clause 62.5).
- The *Project Manager* decides that the *Contractor* has not assessed the event correctly and they do not instruct a revised quotation. The *Project Manager* has the choice of instructing a revised quotation, and they would tend to do so only if they thought that the explanation for requesting a revised quotation would yield the required results.
- If, when the *Contractor* submits required quotations for compensation events, they have not submitted a programme or alterations to a programme that this contract requires. If the programme has changed in any way (e.g. the statement of how the *Contractor* plans to do the work, or a changed Completion Date or a changed Key Date), the *Contractor* must submit a revised programme as part of the compensation event quotation (clause 62.2).
- The *Project Manager* has not accepted the *Contractor's* last programme for a reason stated in the contract by the time the *Contractor* submits the quotation for the compensation event. The *Contractor* is required to submit a revised programme regularly (clause 32.2), but if the latest programme submitted has not been accepted by the *Project Manager* for a reason stated in the contract, then the *Project Manager* is entitled to make their own assessment.

The *Contractor* is therefore incentivised to ensure that their programme is submitted as required. A further incentive exists in clause 64.2, where the *Project Manager* assesses (again, there is no option for the *Project Manager*; the action is obligatory) a compensation event using their own assessment of the programme if there is no Accepted Programme or if the *Contractor* has not submitted alterations or a change to the programme for acceptance as required by the contract.

> The programme is so important that its non-acceptance is grounds for the *Project Manager* making their own assessment of compensation events.

1.9.1.2 Procedure for the Project Manager's assessment

The *Project Manager* has 3 weeks from the time that the need for the *Project Manager's* assessment has become apparent (i.e. 3 weeks from the time the *Project Manager* notified the *Contractor* that they would be making their own assessment) to notify the *Contractor* of their assessment of the compensation event and give them details (clause 64.3). If the *Contractor* was allowed more than 3 weeks to submit their quotation, then the *Project Manager* is allowed that same extra time to do their own assessment and notify the *Contractor*.

1.10. The use of the programme for the assessment of compensation events

Like most contracts, there is a relationship between the cost and time effects of change. In the ECC the programme is part of a quotation for a compensation event, which is a package of time and money.

The programme is therefore intrinsically linked to the effects and management of change.

1.10.1 Programmes that accompany compensation events

The same principles that apply to the Accepted Programme apply to programmes that accompany compensation events. Great care is needed when assessing the time effects where the operation or activity has a learning curve.

For instance, a profile of outputs for a tunnelling contract may look like the example in Figure 1.2.

It can be seen from Figure 1.2 that the point when the compensation event arises and the time when it is to be undertaken may be at different points on the tunnelling progress profile, and so a compensation event issued shortly after work commences in week 1 to lengthen the tunnel by 20 m should be based on the planned output rates in week 10 of 60 m a week and not at the planned output rate of 10 m per week.

Another example is where the *Contractor* has to construct an escalator box with 100 secant piles. The *Contractor* commences work, and they subsequently advise the *Project Manager* that

Figure 1.2 A planned tunnelling progress profile

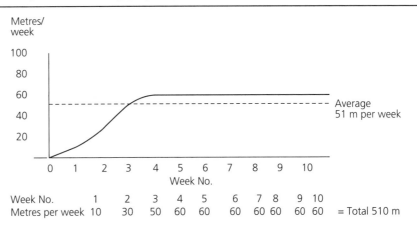

the first 10 piles have taken considerably longer than anticipated due to them encountering physical conditions that an experienced contractor would not have foreseen encountering, namely the ground having physical obstructions. It would be wrong at this stage for the *Project Manager* to assume that the next 90 piles would encounter the same conditions and to award a blanket compensation event, covering all 100 piles. Each pile must be taken on its own merits.

It should also be remembered that the failure of the *Contractor* to supply a programme that demonstrates the effects of a compensation event is grounds for the *Project Manager* to make their own assessment of the time effects of the compensation event.

1.11. Implementation of compensation events

This stage represents the formal conclusion of the administrative process.
Implementation of a compensation event takes place in one of three ways:

- when the *Project Manager* notifies their acceptance of a quotation
- when the *Project Manager* notifies the *Contractor* of their own assessment
- a *Contractor's* quotation is treated as having been accepted by the *Project Manager*.

The third manner of implementation occurs where, under clause 61.4, the *Project Manager* does not reply to a compensation event notification by the *Contractor* within 2 weeks of the notification. In this case, the failure to respond is **treated as acceptance** of the compensation event by the *Project Manager*, as well as an instruction to submit quotations. The *Contractor* is advised to quote this clause with their quotation in this instance.

The acceptance of a quotation is always an alternative to the *Project Manager* doing their own assessment, so the compensation event is implemented when either one occurs, since both cannot happen in the same compensation event.

If

- the *Project Manager* decides as a result of the quotation to give the instruction (or changed decision)

then

- they can either accept the quotation or
- make their own assessment or
- instruct a revised quotation, in which case the revised quotation will either be accepted or assessed by the *Project Manager*.

For all Options, the *Project Manager* notifies the *Contractor* of an assessment made by the *Project Manager* (clause 66.1). When implementing a compensation event, the *Project Manager* includes in their notification implementing a compensation event the changes to the Prices, the Completion Date and the Key Dates accordingly. This will be either from the quotation, which they have accepted, or from their own assessment.

1.12. Reduction of Prices

The only compensation events in clause 60.1 that allow a reduction of the Prices (if the assessment shows a reduction in Defined Cost plus the Fee) are

- a change to the Scope (clause 60.1(1))
- the correction of an assumption made in assessing an earlier compensation event under clause 61.6 (clause 60.1(17)).

The only compensation events in the Option clauses that allow a reduction of the Prices are those arising from clauses 60.4 and 60.6 in main Options B and D and from secondary Option clauses X2.1, X12.3(6) and X12.3(7). All other compensation events listed in clause 60.1 and in the Option clauses cannot lead to reduced Prices even if their effect is to reduce Defined Cost plus the Fee.

1.13 Frequently asked questions
1.13.1 Eight-week time bar

What happens if the *Contractor* notifies the event more than 8 weeks after they became aware of it?

- The first consideration is how to prove when the *Contractor* became aware of the event. The trigger is not when the event occurred but when the *Contractor* became aware of it. In most cases, the *Contractor* should be aware immediately, due to a notification or an instruction or some other visible evidence.
- The second element to consider is what impact the late notification has had on the project and the *Project Manager's* ability to manage the project. A late notification could mean that decisions that had been open to the *Project Manager* are now closed due to passage of time and due to superseding events. It could be said that the *Project Manager* has been prejudiced in their ability to manage the project.
- Contractually, the *Contractor* has 8 weeks **only** to notify a compensation event. If notification does not take place within this period, then the *Contractor* loses their **contractual** right to compensation. They still retain their legal right to compensation, however. This means that an *Adjudicator* should uphold the 8 weeks but a court could instruct compensation.

The *Contractor* should not rely on their legal right to relieve them of their contractual duty, however. There is a reason for the 8-week time bar, and in the interest of the efficacy of the contract and of the mutual trust and co-operation that underpins the contract, the *Contractor* should endeavour to stick to the 8 weeks.

Some clients amend clause 61 to make the 8 weeks a condition precedent for the continuation of the procedure and the entitlement to an assessment of time and money. It could well be argued, however, that such a clause is against the rules of natural justice and justified enrichment. This could be particularly applicable where the *Project Manager* has not carried out their actions under the contract, such as notified a compensation event resulting from their own actions. If the *Contractor* does not notify a compensation event within 8 weeks of becoming aware of it, and the event is a matter that should have been notified to the *Project Manager* as a compensation event but was not, then the *Contractor* does not lose their entitlement to an assessment of time and money. Clause 61.7 states that a compensation event is not notified after the issue of the Defects Certificate. It does not seem to be in the spirit of the NEC to wait until the last minute to notify a quotation, however (see the final point in Section 1.13.2 below).

In conclusion, it is recommended that all the circumstances should be taken into consideration prior to not accepting a compensation event notification. Above all else, mutual trust and co-operation as embodied by fair and reasonable actions is the philosophy of the NEC, and to act otherwise could result in unnecessary conflict.

1.13.2 Claims

What happens if the *Contractor* submits a 'claim'-type document?

- Where the *Contractor's* compensation event notification was accepted and they were instructed to submit quotations, a poorly drafted quotation does not detract from the fact that the compensation event has been validly notified and accepted.
- The *Project Manager* may choose to make their own assessment; however, this should always be a last resort since not only is it difficult to do but it may result in a dispute.
- The most effective action to take would be to sit with the *Contractor* and go through the *Project Manager's* expectations of a quotation. In particular, it should be explained that global forecasts are not acceptable and that quotations cannot be revisited after the compensation event has been implemented.
- The incidence of claims may increase since the *Contractor* is no longer obliged to notify a compensation event if the *Project Manager* does not. Because there is no time limit on the notification of a compensation event in this case (other than that it must take place before the *defects date*), the *Contractor* may notify the event at Completion, although the event took place much earlier. This is not in the spirit of the NEC, however, nor does it help the *Contractor's* cash flow, and they are advised to notify the event themselves.

Managing Change

1.13.3 The *Project Manager* does not notify

What happens if the *Project Manager* does not notify a compensation event that they should have done?

- The principle in the NEC is that where the *Client*, the *Project Manager* or the *Supervisor* does not carry out their obligations, it is a compensation event. Where the *Contractor* is in breach of contract, there is no contractual remedy. This is based on the premise that the *Contractor* is likely to be financially influenced by their own breaches and is therefore less likely to commit them in the first place.
- In this case, however, there is no direct sanction on the *Project Manager* for not notifying a compensation event that they should notify. This is possibly because of the inherent failsafe that the *Contractor* may notify the event if the *Project Manager* does not. It is therefore in the *Contractor's* interests to notify compensation events.

1.13.4 Early warnings

How does an early warning affect a later compensation event on the same matter?

- The *Contractor* and the *Project Manager* are both obliged to notify an early warning as soon as either becomes aware of any matter that could
  - increase the total of the Prices
  - delay Completion
  - delay meeting a Key Date or
  - impair the performance of the *works* in use.

  This is to give the *Contractor* and the *Project Manager* time to consider the implications of the matter and to take action to mitigate any potential consequences.
- If the *Project Manager* decides that the *Contractor* did not notify an early warning that an experienced contractor could have notified and the same matter becomes a compensation event, the *Project Manager* informs the *Contractor* of this decision when they instruct the *Contractor* to submit quotations (clause 61.5).
- Notifying the *Contractor* in this way means that the *Contractor* has to assess the compensation event as if they had given the early warning (clause 63.7), and it means that the *Project Manager* may assess the compensation event in the same way if they have chosen to assess the event themselves.
- The reason for this procedure is to ensure that the *Contractor's* not notifying an early warning matter does not prejudice the *Project Manager* in their management of the project. If, for example, the matter had been identified, avenues available to the *Project Manager* at that time might have been sufficiently flexible to facilitate the most economical route to have been chosen.

1.13.5 Grouping compensation events

What does the *Project Manager* do if many compensation events take place over a short period of time? Do they have to attend to each separately?

Because the compensation event procedure is fairly long and complex, it is understood that using the procedure for small compensation events that will be carried out in a few hours seems a little arduous.

Many *Project Managers* allow the grouping of smaller compensation events into one notified compensation event on a specified day of the week, for example Friday. All the smaller events that took place during the event are then collected and notified in one compensation event notification. Larger compensation events that require time and effort to assess are still notified separately.

This should only affect Option C and D contracts. Option A and B contracts should not be subject to many compensation events due to the philosophy behind the fixed price required. Compensation events under Option E are mostly important for time purposes rather than budget purposes, although the compensation event quotations are obviously included in the forecast of Defined Cost (budget).

Another, related, matter is where *Contractors* have not correctly forecast the consequential results of a compensation event. This could happen where a relatively minor compensation event of low value has a large consequential impact on the programme. Many *Project Managers* in this situation allow a dummy compensation event that sweeps up the consequential events of previous compensation events.

1.13.6 Amending the contract prior to execution

What does the *Project Manager* do if the drawings change before they have issued the contract? Can they use the compensation event procedure to change the prices to reflect the new drawings?

The compensation event procedure is a part of the contract, and since the contract in this situation is not yet in place, it is a little incongruous to be using a contractual procedure to sort out something that is happening before the contract. You can do virtually anything by agreement, however, and if both Parties agree to use the tendered data for the SCC to amend the tendered prices to reflect updated drawings, then this is acceptable. It is probably difficult to check the validity of the quotation, however, given the lack of documentation accompanying a previous event.

More importantly, however, is that fact that the Prices are changed before the contract is executed. As long as you have the time available to make these changes, this is better than issuing an obsolete contract and immediately issuing a score of compensation events.

1.13.7 Removing compensation events

Of the 21 compensation events contained in clause 60.1, the two that are most frequently varied are clauses 60.1(12) and 60.1(13); that is, the compensation events that deal with physical conditions and weather.

The basic premise of a compensation event is that the *Client* takes the risk for the event described. In this way, the risks to be taken by the *Contractor* are clearly laid out, and the *Contractor* is able to price the contract effectively, taking into account those elements of the contract that are at their risk. Many clients are so used to writing out ground conditions in an Institution of Civil Engineers (ICE) contract that they automatically want to exclude clause 60.1(12) of the ECC as well. It may therefore be worth spending just a little time on the concept of risk and how it affects the contract.

> Risk is not transferred but is reallocated.

One of the principal misconceptions about risk is that it is 'transferable'. Some clients like to use the phrase 'transfer the risk to the *Contractor*'. In general, however, the risk is not 'passed' to the *Contractor*, but rather it takes on a different form and is reallocated. If, for example, in a traditional contract that is not the ECC, the *Client* decides they do not want to take the risk of ground conditions and they rewrite the contract so that the *Contractor* has the risk of ground conditions, the risk has assumed a different form for the *Client*.

If the contract is a lump sum contract, it is likely that the *Contractor* will build the potential cost of such a risk into the contract price. They might be conservative in their estimate of the occurrence of the risk and so may include a monetary value in the contract price at a high level to cover their risk. This might not even stop them submitting a claim to the *Project Manager* if the risk does materialise and the cost to the *Contractor* is far in excess of that included in the contract price. If, on the other hand, the risk does not materialise, all things being equal, the *Contractor* will have pocketed the cost of the risk. Whether the risk occurs or not, therefore, the *Client* will pay.

In a cost-reimbursable-type contract, the *Client* will pay for the risk whether it is allocated to the *Contractor* or to the *Client*. It is unlikely that the *Contractor* will accept a risk unconditionally, and perhaps the *Client* should consider whether it is fair and reasonable that the *Contractor* takes the risk for something that is unforeseen and outwith the *Contractor's* control. In general, therefore, it is more effective for the *Client* to pay for a risk that does occur, than to pay for something that might happen.

Getting back to the ECC and compensation events, it is worth noting that all the events are well defined. In particular, the two most contentious events, namely physical conditions and

Managing Change

weather, are more than simply 'unforeseen ground conditions' or 'inclement weather'. The approach is more objective, and therefore more measurable than in traditional contracts.

1.13.8 Adding compensation events

The compensation events contained in the main and secondary Options are optional by virtue of their being part of an Option, and therefore these will not be discussed further.

The *Client* can modify the contract risk allocation by identifying in the Contract Data additional compensation events In adding an element, the *client* should consider the following:

- What is it that will affect the project? Is it gusts or a constant wind speed above a certain speed?
- What does the Met Office measure and how near to the Site is this measured?
- Does the *Client* want to take the risk for this? That is, what will be the effect on the project and who is best placed to manage this risk?
- What is the policy for managing situations where there is no fault? (This is helped to a certain extent by compensation event 60.1(19).)

> Risk should be allocated to the Party best placed to manage it.

Consider the case where high-speed gusts of 120 mph blow down a structure, but wind is not an additional *weather measurement*. Since wind is not a *weather measurement*, there can be no compensation event, but it is advisable that the *Project Manager* considers the situation anyway and perhaps comes to a commercial settlement. For example, were the winds forecast and did the *Contractor* take what precautions they could in the time allowed to prevent damage? Did the *Client* facilitate the *Contractor's* actions? What is the extent of the damage to the project and how will this event affect the rest of the project? It may be in the interests of the project to provide the *Contractor* with the means to very quickly repair the damage so that a critical date

Table 1.2 ECC clauses where the *Project Manager* fails to act

Clause	Description
31.3	'If the *Project Manager* does not notify acceptance or non-acceptance within the time allowed, the *Contractor* may notify the *Project Manager* of that failure. If the failure continues for a further one week after the *Contractor's* notification, it is treated as acceptance by the *Project Manager* of the programme.'
61.3	'The *Project Manager* has not notified the event to the *Contractor*.'
61.4	'If the *Project Manager* fails to reply to the *Contractor's* notification of a compensation event within the time allowed, the *Contractor* may notify the *Project Manager* of that failure. If the failure continues for a further two weeks after the *Contractor's* notification it is treated as acceptance by the *Project Manager* that the event is a compensation event and an instruction to submit a quotation.'
62.6	'If the *Project Manager* does not reply to a quotation within the time allowed, the *Contractor* may notify the *Project Manager* of that failure. If the *Contractor* submitted more than one quotation for the compensation event, the notification states which quotation the *Contractor* proposes is to be used. If the failure continues for a further two weeks after the *Contractor's* notification it is treated as acceptance of the quotation by the *Project Manager*.'
64.4	'If the *Project Manager* does not assess a compensation event within the time allowed, the *Contractor* may notify the *Project Manager* of that failure. If the *Contractor* submitted more than one quotation for the compensation event, the notification states which quotation the *Contractor* proposes is to be used. If the failure continues for a further 2 weeks after the *Contractor's* notification, it is treated as acceptance by the *Project Manager* of the quotation.'

Compensation events

1.13.9 The *Project Manager* fails to act

is met. Future relations with the *Contractor*, social and environmental responsibilities, and insurances should all also be considered by the *Project Manager*.

Failure of the *Project Manager* to act (Table 1.2) is one of the most frustrating and difficult things for a *Contractor* to contend with. There is no simple remedy to this problem other than the hope that the *Client* will employ the right competency of person and that they will recognise when their *Project Manager* is failing to act, and if this is persistent then they will replace them.

The contract includes sanctions for non-performance of the *Project Manager*. A *Project Manager* who fails to perform may find themselves subject to a claim on their professional indemnity insurance from the *Client*.

Clause W1.1(4) states that the *Client* may take a treated as accepted quotation to adjudication.

1.14. Format of a compensation event quotation

A quotation for a compensation event comprises

- proposed changes to the Prices
- any delay to the Completion Date and Key Dates assessed by the *Contractor*.

See Appendix 2 of Chapter 2 in this book for an example of a compensation event quotation. See Chapter 2 for how to use the SCC. Appendix 2 of Chapter 2 contains example standard forms that may be used during the compensation event procedure.

Managing Change
ISBN 978-0-7277-6188-0

ICE Publishing: All rights reserved
http://dx.doi.org/10.1680/mc.61880.035

Appendix 1
Compensation event procedure

Synopsis   This appendix describes the compensation event procedure through

- listing actions to be taken by the *Contractor* and *Project Manager*
- flow-charting the procedure
- drafting a timeline for the compensation procedure notified by both the *Contractor* and the *Project Manager*.

Section A: Actions required to be taken by the *Project Manager*

Key
PM *Project Manager*
C *Contractor*
CE compensation event
S *Supervisor*

A1.1. Notifications

Clause	Action
61.1	The *PM* **notifies** the *C* of a CE arising from the *PM* or *S* giving an instruction or notification, issuing a certificate or changing an earlier decision.
61.4	The *PM* **notifies** their decision to the *C* when the *PM* consider that an event notified by the *C* is not a CE or the *PM* notifies the *C* when the *PM* decides otherwise.
61.5	The *PM* **notifies** their decision to the *C* that they believe that the *C* did not give an early warning of the CE that an experienced *C* could have given.
61.6	The *PM* **notifies** corrections to earlier assumptions upon which they instructed the *C* to base their quotation in cases where effects of the CE were too uncertain at the time of the event.
62.3	Following submission by the *C* of a quotation for a CE, the *PM* may **notify** the *C* that the *PM* will be making their own assessment (for reasons stated in clause 64).
65.3	Where the *PM* has instructed the *C* to submit a quotation for a **proposed** instruction, following submission of the quotation, the *PM* may **notify** the *C* that the proposed instruction will not be given.
62.5	The *PM* **notifies** the *C* of any extensions to the time allowed for either the *C* to submit quotations or the *PM* to reply to quotations.
64.3	The *PM* **notifies** the *C* of their assessment of a CE and gives the *C* details of the *PM's* own assessment.
66.1	The *PM* implements a CE by **notifying** the *C* of acceptance of the *C's* quotation. The *PM* **notifies** the *C* of an assessment made by the *PM* or the *C's* quotation is treated as having been accepted by the *PM*.

A1.2. Instructions

Clause	Action
61.2	The *PM* **instructs** the *C* to submit quotations for CEs arising from the *PM* or *S* giving an instruction or changing an earlier decision.
65.1	The *PM* may **instruct** the *C* to submit quotations for a proposed instruction or a proposed changed decision.
61.4	The *PM* **instructs** the *C* to submit a quotation when the *PM* decides that an event notified by the *C* is a CE.
62.1	The *PM* may **instruct** the *C* to submit alternative quotations based upon different ways of dealing with a CE.
62.4	Following submission by the *C* of a quotation for a CE, the *PM* **instructs** the *C* to submit a revised quotation.

A1.3. Acceptances

Clause	Action
61.4	If the *PM* does not reply within 2 weeks of the *C's* notification of a CE, then the notification is treated as being accepted by the *PM*.
62.2	Following submission by the *C* of a quotation for a CE, the *PM* **accepts** the quotation.

Section B: Actions required to be taken by the *Contractor*

Key
PM   *Project Manager*
C   *Contractor*
CE   compensation event

B1.1. Notifications

Clause	Action
61.3	For CEs not notified by the *PM*, the *C* **notifies** the *PM* that they believe the event is a CE and the *PM* has not **notified** the event to the *C*.
61.4	The *C* may notify the *PM* if the *PM* has not responded timeously to a CE notification from the *C*.

B1.2. Submissions

Clause	Action
62.3	The *C* **submits** quotations for CEs.
62.4	The *C* **submits** revised quotations for CEs.
64.4	The *C* may notify the *PM* if the *PM* has not assessed a CE within the time allowed.

Section C: Compensation event procedure

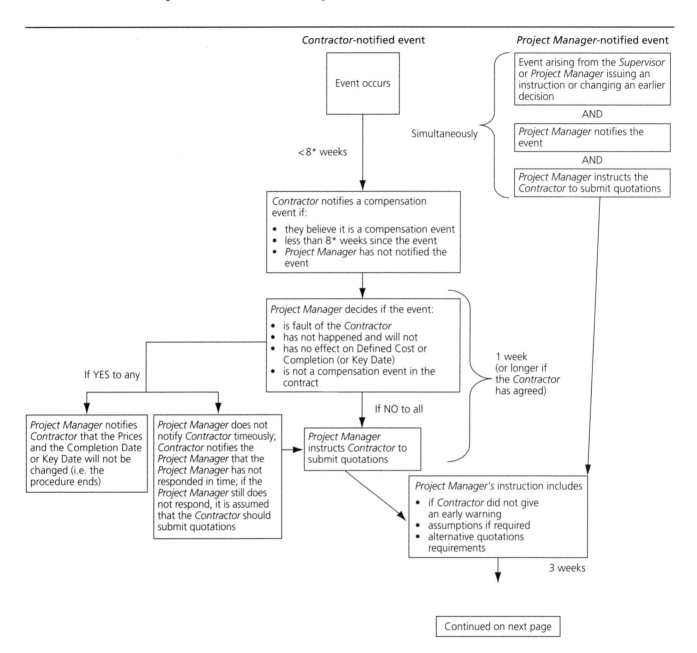

Appendix 1

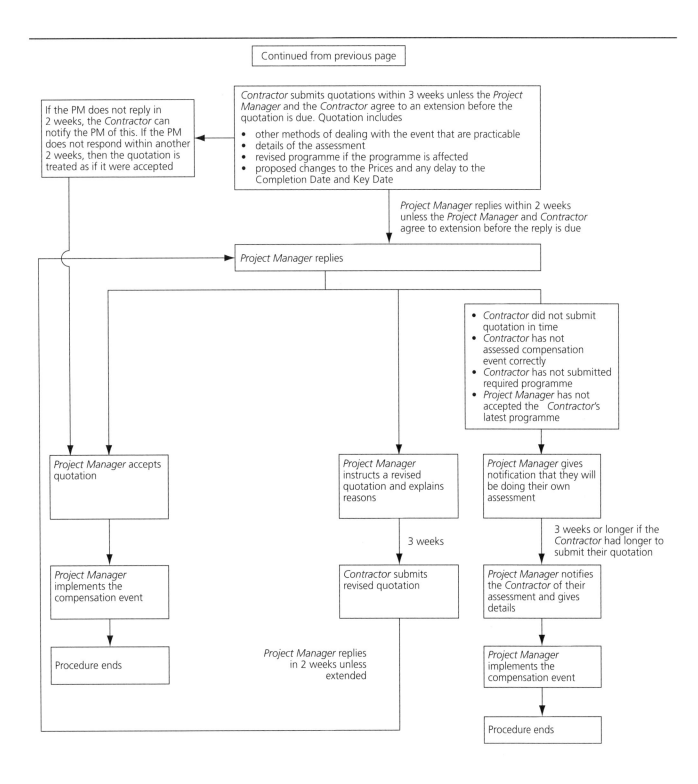

Managing Change

Section D: Procedure for events notified by the *Project Manager*

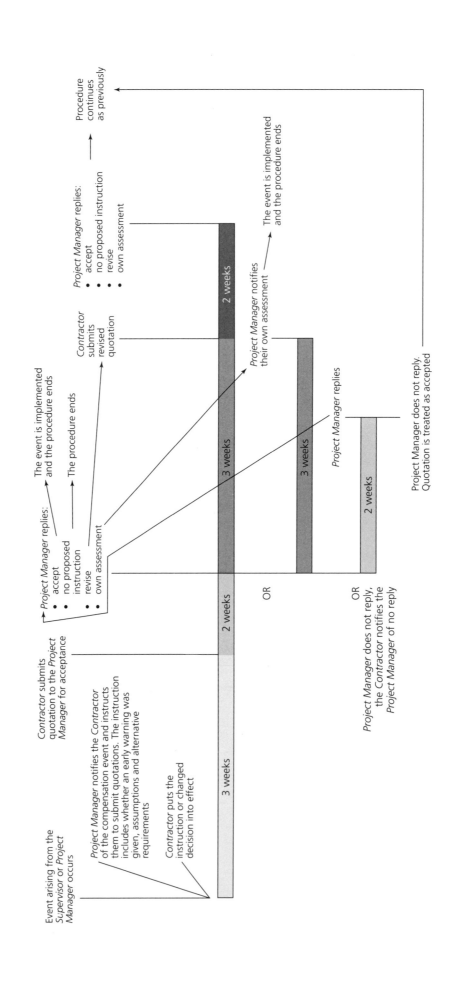

Appendix 1

Section E: Procedure for events notified by the Contractor

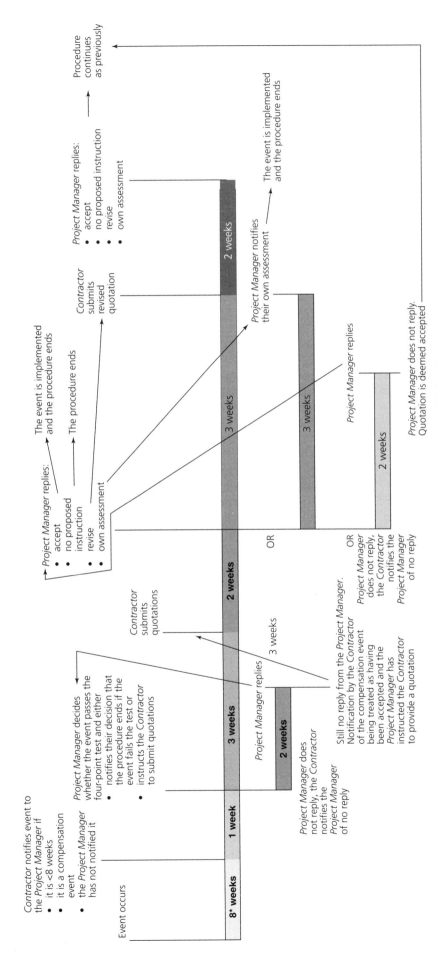

* ECC states that if the Contractor does not notify an event within 8 weeks of becoming aware of the event, they are not entitled to changes in the Prices, Completion Date or Key Date unless the Project Manager should have notified the event to the Contractor but did not.

Section F: Most common complicating factors

Key
C   *Contractor*
PM   *Project Manager*
SRs *Senior Representatives*
CE   compensation event

Complicating factor	Action required	Further possible action
The *PM* decides that an event notified by the *C* as a CE is not a CE (clause 61.4)	The *C* may notify the *PM* of their intention to submit a dispute to the *SRs* (clause W1.1)	The *PM* within 2 weeks of the *C's* notification changes the decision previously communicated to the *C* and instructs the *C* to submit a quotation (clauses 60.1(8) and 61.4) OR The *PM* does not change the decision previously communicated to the *C* and risks the *C* submitting a dispute to the *SRs*
The *C* fails to submit a quotation and details of their assessment within 3 weeks or other extended period agreed (clause 62.3)	The *PM* assesses the CE and notifies the *C* of their assessment within the same period originally allowed to the *C* for submission (clauses 64.1 and 64.3)	The *C* may notify the *PM* of their intention to submit a dispute to the *SRs* if they believe the *PM's* assessment to be incorrectly calculated (clause W1.1 or W2.1)
The *PM* believes that the *C* has not assessed the CE correctly in a quotation submitted	The *PM* instructs the *C* to submit a revised quotation, explaining their reasons for doing so (clause 62.4) OR The *PM* assesses the CE and notifies the *C* of their assessment (clauses 64.1 and 64.3)	The *C* resubmits the quotation (clause 62.4) OR The *C* may notify the *PM* of their intention to submit a dispute to the *SRs* if they believe the *PM's* reasons to be ill judged or are not for a reason stated in the contract (clause W1.1 or W2.1) OR The *C* may notify the *PM* of their intention to submit a dispute to the *SRs* if they believe the *PM's* assessment to be incorrectly calculated (clause W1.1 or W2.1)

Managing Change
ISBN 978-0-7277-6188-0

ICE Publishing: All rights reserved
http://dx.doi.org/10.1680/mc.61880.043

Chapter 2
Schedule of Cost Components and Short Schedule of Cost Components

Synopsis   This chapter discusses aspects relating to the full SCC and its short version the SSCC, including

- when the SCC and SSCC are used
- how the SCC and SSCC interact with the payment clauses
- Defined Cost
- the Fee
- the components of cost included under the SCC and SSCC
- Contract Data part two.

2.1. Introduction

The ECC, in common with other standard forms of construction contracts, provides rules for assessing

- the amount to be paid to the *Contractor* for work done
- the change to amounts to be paid to the *Contractor* for variations and other events that may arise and that, under the contract, are the *Client's* risk, collectively termed compensation events by the ECC.

The SCC is used for main Options C, D and E, whereas the SSCC is used for main Options A and B.

Given, then, the obvious connection of the SCC (and its short version) to the financial outcome of a contract it is perhaps unfortunate that the SCC has so far proven to be one of the least understood parts of the ECC. This chapter therefore sets out to examine the purpose and detail of the SCC using examples and discussion to provide a better understanding of this fundamentally important part of the ECC.

2.2. What is the SCC?

To understand the role of the SCC in the administration of the contract we need to first identify where it is referred to in the contract.

A search reveals that the only reference to it occurs in the definition of Defined Cost. Since the definition of Defined Cost differs depending on which main Option of the ECC is used, it is to the main Option clauses that we must turn.

It has been stated that the principal role of the main Option clauses is to determine how the *Contractor* is to be reimbursed for their efforts or, expressed more formally, where the financial risk boundary is set between the *Client* and *Contractor* (refer to Chapter 2 of Book Two). This immediately establishes that the role of the SCC must be linked to how the *Contractor* is reimbursed.

To confuse matters a little in this area there are in fact two SCCs, namely

- the full schedule – the SCC
- the short schedule – the SSCC.

What follows relates to the SCC, since an understanding of this will make it far easier to understand the differences between the use and content of the two versions.

The SCC is a complete identification of the components of cost for which the *Contractor* will be reimbursed under certain circumstances. These components of cost are a part of Defined Cost as defined.

It is important to note that the definition of Defined Cost differs, depending upon which of the six main Options A to F is being used.

These components of cost are **not priced at the time of tender**. However, in Contract Data part two the *Contractor* is required to insert certain information in relation to these components of cost.

> The SCC is a complete identification of the components of cost. The definition of Defined Cost does not mean all of the *Contractor's* costs, but is confined to the components as listed in the SCC.
>
> The definition of Defined Cost differs, depending upon which of the six main Options you are using.
>
> Contract Data part two contains information required for use in conjunction with the SCC. It is therefore **very important** that this is completed correctly at the time of tender.

2.3. Why has this approach been taken?

Traditionally, the valuation of change has been assessed on the basis of tendered rates and prices. Problems occur, however, when, as often happens on projects, the scope and nature of the project start to vary, and arguments then arise with regard to the applicability of bills of quantities rates, prices and lump sum items and how much the quantity/type/scope of an item needs to change before a new rate or price is required.

It is possible to argue forever about the rights and wrongs of a particular price. The SCC is a way around these problems.

The ECC promotes the idea of the pre-assessment of change via a quotation. This supports the concept of 'stimulus to good management'. A well-run project will identify change at the earliest time possible. This supports the idea that 'foresight applied collaboratively mitigates problems and shrinks risk'.

In the ECC, all change is valued at 'Defined Cost', with no reference made to tendered rates or prices. If the *Project Manager* and *Contractor* agree, rates and lump sums can be used to assess a compensation event for all main Options (clause 63.2) The philosophy behind this provision is that the *Contractor* should be 'no better nor no worse off' as a result of change that is at the risk of the *Client* under the contract during the construction of the *works*.

This approach also enables the *Client* to call for quotations for a number of options on how to deal with the change. Therefore, if the *Client* has a facility to be opened by a certain date or costs are of paramount importance, then the *Client* can consider and the *Project Manager* can instruct the *Contractor* to submit quotations for these options.

The *Contractor* is conceptually in the same position when pricing the quotation as they would have been at the time of tender. They also, as with tendering, carry the risk should their quotation be wrong.

> The idea with the SCC and SSCC is that conceptually the *Contractor* is in the same position for a compensation event as when they tender for the work.

2.4. Assessment options

The default situation for assessing the changes to the Prices in Options A and B is the SSCC. However, there is an option to use rates and lump sums if the *Project Manager* and *Contractor* agree (clause 63.2).

The default situation for assessing the changes to the Prices in Options C, D and E is the SCC.

2.5. When is the SCC or SSCC used?

2.5.1 Priced-based contracts – Option A and B contracts

Options A and B are the priced-based Options (lump sum and remeasurable, respectively). The definition of the Price for Work Done to Date under these options is not based on Defined Cost but on, respectively, the Activity Schedule and the Bill of Quantities (i.e. the *activity schedule* and the *bill of quantities* for Options A and B). Therefore, under these main Options the SSCC is used to evaluate compensation events only.

Option A – payment during the contract

For Option A contracts, the *Contractor* is reimbursed using the Activity Schedule for their work carried out during the period of the contract and priced at the tender stage. That is, the *Contractor* is paid the lump sum price listed in the Activity Schedule when that activity is completed. Each completed activity then forms part of what is termed the Price for Work Done to Date (clause 11.2(29)).

Option A – evaluating compensation events

The financial effects of all compensation events that occur during the period of the contract are not priced using lump sum prices in the Activity Schedule as a basis, but using the SSCC. That is, any additional costs arising under the contract, including additional Subcontractor costs, are quoted for using the SSCC. However, there is a facility to use lump sums and rates to assess compensation events instead of Defined Cost (clause 63.2).

The accepted compensation events are added to the lump sum Prices in the Activity Schedule, and the *Contractor* is therefore paid for the compensation events through the mechanism of the Price for Work Done to Date.

> **Option A**
>
> The lump sum prices inserted by the *Contractor* at the tender stage for each activity on the Activity Schedule are not necessarily used to assess the financial effect of compensation events.
>
> The SSCC is only used for the assessment of change to the Prices as a result of compensation events.
>
> Tendered rates and prices are **not used** to assess change unless agreed between the *Contractor* and the *Project Manager* (clause 63.2).

Option B – payment during the contract
For Option B contracts, the *Contractor* is reimbursed using the Bill of Quantities for their work carried out during the period of the contract and priced at tender stage.

Option B – evaluating compensation events
The financial effects of compensation events that occur during the period of the contract are not priced using the rates and prices in the Bill of Quantities as a basis, but using the SSCC. This is the default position described in the contract, and may discourage possible front loading of the bill rates and prices impacting on costs. However, for Option B contracts, the *Project Manager* and the *Contractor* can agree to use lump sums and rates, including those in the *bills of quantities* instead of the SSCC, to calculate the cost of compensation events (clause 63.2).

> **Option B**
>
> The rates, prices and lump sums inserted by the *Contractor* at the tender stage for each Bill of Quantities item are not used to assess the financial effects of compensation events.
>
> By agreement between the *Project Manager* and *Contractor*, the rates or lump sums, including those in the *bills of quantities*, can be used to assess the financial effects of compensation events (clause 63.2).

2.5.2 Cost-based contracts – Option C, D and E contracts

In addition to the use of the SCC for assessing the financial effect of compensation events under Option C, D and E contracts, it is also used as the basis for calculating the amount due to the *Contractor*.

Options C and D – target cost
The Activity Schedule (for Option C) or the Bill of Quantities (for Option D) that is priced by the *Contractor* at the tender stage is included in the contract only to provide the target cost. This target cost may be changed through compensation events ('in accordance with this contract'). The Prices are only considered during the contract

- to account for changes through compensation events to ensure that the base comparison of the target cost remains realistic
- as a comparison with the Price for Work Done to Date to assess the *Contractor's* share.

The Prices are not used to pay the *Contractor* during the period of the contract, and this definition will therefore not be considered further in this chapter. For a fuller discussion of target cost and the *Contractor's* share, see Chapter 2 of Book Two on contract options.

Options C, D and E – payment during the contract and the evaluation of compensation events
The Price for Work Done to Date is effectively the amount due to the *Contractor* as assessed by the *Project Manager*. The Price for Work Done to Date is therefore the amount paid to the *Contractor* during the period of the contract. The Price for Work Done to Date is not related

Schedule of Cost Components and Short Schedule of Cost Components

to the Prices except where the Prices are used as a comparison against the Price for Work Done to Date to determine the *Contractor's* share (Options C and D).

The SCC is therefore the only method for the *Contractor* to be reimbursed their costs under Option C, D or E contracts.

The Price for Work Done to Date comprises a number of elements, as shown in Figure 2.1.

Figure 2.1 Example of the amount due for main Options C, D and E

Price for Work Done to Date (PWDD)	Plus other amounts to be paid or retained	
Clause 11.2(31): total Defined Cost that the *Project Manager* forecasts will have been paid by the *Contractor* before the next assessment date plus the Fee	**Plus amounts to be paid to the Contractor**	**Less amounts to be retained from the Contractor**
		Clause 25.2: the *Contractor* does not provide services and other things as stated in the Scope
		Clause 25.3: work does not meet a Condition for a Key Date
		Clause 41.6: costs incurred in repeating a test or inspection
		Clause 50.5: no programme identified in the Contract Data
	Clause 51.2: interest on late payment	
	Clause 51.3: interest on the correcting amount	
	Clause 51.5: any tax that the law requires	
		Clause 86.1: the *Client* insures a risk that the contract requires the *Contractor* to insure
	Clause 86.3: the *Contractor* insures a risk that the contract requires the *Client* to insure	
	Option X1: price adjustment for inflation (only used with Options A, B, C and D)	Option X1: price adjustment for inflation (only used with Options A, B, C and D)
	Clause X3.2: multiple currencies excess (only used with Options A and B)	
	Option X6: bonus for early Completion	
	Clause X7: delay damages	
	Clause X12.4(1): partnering incentives	
		Option X14: advance payment – repayment
	Option X16: retention (not used with Option F)	Option X16: retention (not used with Option F)
	Option X20: Key Performance Indicators (not used with Option X20)	Option X20: Key Performance Indicators (not used with Option X20)
	Option X22: early *Contractor* involvement (used only with Options C and E) Clause X22.8: incentive payment	
	Options C and D	
	Contractor's share at Completion Prices < PWDD	*Contractor's* share at Completion Prices > PWDD

> **Options C and D**
>
> The lump sum prices inserted by the *Contractor* at the tender stage for each activity on the Activity Schedule or Bill of Quantities are not used to assess the financial effects of compensation events.
>
> The SCC is used to assess compensation events and to determine the Price for Work Done to Date.
>
> Where agreed, rates and lumps may be used to assess the financial effects of compensation events.
>
> Tendered rates and prices are **not used** to assess the financial effects of compensation events.
>
> **Option E**
>
> The SCC is used to assess both the financial effects of compensation events and to determine the Price for Work Done to Date by the *Contractor*.

2.5.3 Option F contracts

The SCC is not used at all in Option F contracts since the management *Contractor* is paid the amounts due to Subcontractors. The assumption is that the management *Contractor* does not do any of the work themselves and therefore they do not need to be reimbursed in a manner such as that described by the SCC. Where the *Contractor* does do work, they are paid the lump sum stated in Contract Data part two.

2.5.4 SCC summary

Table 2.1 summarises the use of the SCC and the SSCC for the six main Options.

Table 2.1 Uses of the SCC and SSCC

Main Option	Contract type	Evaluation of the financial effects of compensation events	Payment
A	Priced based	SSCC only; or rates and lump sums	Completed activities in the Activity Schedule
B		SSCC only; or rates and lump sums	Bills of Quantities items multiplied by the rate
C	Cost based	SCC or rates and lump sums	SCC
D		SCC or rates and lump sums	SCC
E		SCC or rates and lump sums	SCC
F	Management contract	SCC does not apply	Amount of payments to Subcontractors (and the *prices* for work done by the *Contractor* themselves)

2.6. Defined Cost

In each of the main Options, a clause exists that provides a definition of Defined Cost. A summary of the definitions is given in Table 2.2.

From Table 2.2 it can be seen that Defined Cost is defined by the SCC, which provides a list of the components of cost to which the *Contractor* is entitled. This includes such items as wages and salaries and the listed components of the cost of Equipment and Plant and Materials. This is different from some other conditions of contract, where 'actual cost' is used but not defined clearly and objectively.

Table 2.2 Definition of Defined Cost for the six main Options

Main Option	Clause	Definition of Defined Cost
A	11.2(23)	Defined Cost is the cost of the components in the Short SCC
B	11.2(23)	As Option A
C	11.2(24)	Defined Cost is the cost of the components in the SCC less Disallowed Cost (clause 11.2(26))
D	11.2(24)	As Option C
E	11.2(24)	As Option C
F	11.2(25)	Defined Cost is the amount of payments due to Subcontractors for work that is subcontracted without taking account of amounts paid to or retained from the Subcontractor by the *Contractor* that would result in the *Client* paying or retaining the amount twice and the *prices* for work done by the *Contractor* themselves less Disallowed Cost

Note that Defined Cost is used only to calculate changes in main Options A and B, but used for all payments in Options C, D and E

> The definition of Defined Cost differs depending upon which of the main Options you are using. You need to look at the particular main Option clauses to determine how to assess changes and how to assess the amount due.

2.7. The Fee

All the costs to the *Contractor* not covered under the SCC are deemed to be covered by the Fee (clause 52.1). The Fee is calculated by multiplying the *fee percentage* tendered by the *Contractor* in Contract Data part two with their correlating Defined Cost, which is the total of the components of cost as listed in the SCC:

Fee = *fee percentage* × Defined Cost of subcontracted work

The *fee percentage* should therefore cover

- the *Contractor's* desired profit
- overheads that cannot be claimed as a component of cost in the SCC, meaning head office overheads, since 'site overheads' or 'preliminaries' are generally covered in the SCC.

Examples of inclusions in the *fee percentage* are

- profit
- head office charges and overheads
- corporation tax, insurance premiums (it should be noted that *Client's* liability insurance comes under the people cost component in the SCC)
- advertising and recruitment costs
- sureties and guarantees for the contract.

2.7.1 The Fee – defined term

The Fee (clause 11.2(10)) is the sum of the amounts calculated by applying the *fee percentage* to the Defined Cost of the work.

2.8. The components of cost included under the SCC

The SCC only applies to the main Option C, D and E cost-based contracts and defines the cost components for which the *Contractor* will be reimbursed included in an assessment of changed costs arising from a compensation event (C, D and E).

As a general policy, the SCC only provides for the direct reimbursement of those cost components that are readily identifiable. The last sentence of the opening paragraph to the SCC states that: 'An amount is included only in one cost component and only if it is incurred in order to Provide the Works.' The principal way that this is achieved is through the concept of Working Areas. Working Areas under the ECC mean the areas of land comprising the Site as made available by the *Client*, together with any additional areas proposed by the *Contractor* (in Contract Data part two and through clause 16.3) and accepted by the *Project Manager* as being necessary to Provide the Works.

Generally, it is only the cost of People and Equipment working within these Working Areas that is reimbursed as Defined Cost. This recognises the difficulty and therefore the higher risk to the *Client* in identifying and controlling costs that are incurred away from the Site; that is, outside the Working Areas.

The exception to this general principle is those costs incurred by the *Contractor* associated with the design, manufacture and fabrication outside of the Working Areas that are dealt with in separate cost components 6 and 7 in the SCC.

To make the SCC fully effective requires certain data to be inserted by the *Contractor* in Contract Data part two. It is important that all data required by Contract Data part two for the SCC is inserted by the *Contractor* **before submitting their tender**, since the decision to use one or the other arises during the contract by agreement between the *Contractor* and the *Project Manager* rather than before the contract starts.

There are eight headings for the components of cost included in the SCC:

1 'People'
2 'Equipment'
3 'Plant and Materials'
4 'Subcontractors'
5 'Charges'
6 'Manufacture and fabrication'
7 'Design'
8 'Insurance'.

Under each of these headings are described all the components of cost for which the *Contractor* will be reimbursed during the period of the contract and that will also be taken into consideration when evaluating the financial effects of compensation events. Any other components of cost not identified in the definition of Defined Cost (and therefore not included in the SCC) are deemed to be included in the Fee (clause 52.1). *Contractors* should therefore ensure that the tendered fee percentage included in their Contract Data part two covers all elements of cost not included in the SCC.

2.8.1 Working Areas

Note that except for cost components 6 and 7, the components in the SSCC are all for costs within the Working Areas. This means that the identification of the *working areas* by the *Contractor* in Contract Data part two is particularly important. The *working areas* are generally identified as the Site and other areas adjacent or near to the Site that the *Contractor* considers they might use temporarily for the purpose of Providing the Works. Examples are 'borrow pits' or a concrete batching facility. These should be identified as *working areas*.

A problem often raised by *Contractors* is that at the tender stage they do not know precisely where, say, a 'borrow pit' will be, as they are still in negotiation with land owners and the like. In these circumstances, it is important for the *Contractor* to identify in the Contract Data that they intend to have such a location, even though they may only be able to provide a generic description without a precise location.

2.8.2 Cost component 1 – people

The cost components for people in the SCC comprise the following categories:

- Items 11, 12 and 13 for people directly employed by the *Contractor* and whose normal place of working is within the Working Areas; that is, direct employees of the *Contractor* whose normal place of working is within the Working Areas (e.g. tradesmen and the site agent).
- Items 11, 12 and 13 for people directly employed by the *Contractor* and whose normal place of working is **not** within the Working Areas but who are working in the Working Areas proportionate to the time they spend working in the Working Areas; that is, direct employees of the *Contractor* whose normal place of working is not within the Working Areas but who are working in the Working Areas (e.g. a specialised tradesman who is being used for a compensation event).

Item 14 includes a third category of cost for people and only one description of cost to be included in the SCC and therefore payment to the *Contractor*:

'The following components of the cost of people who are not directly employed by the *Contractor* but are paid for by the *Contractor* according to the time worked while they are within the Working Areas.

Amounts paid by the *Contractor*.'

The words 'Amounts paid by the *Contractor*' may lead to the assumption that whatever the *Contractor* pays, the *Client* is liable to pay. However, one needs to remember that in the SCC this only applies to main Options C, D and E, and that this cost component will therefore come under the scrutiny of Option D–E Clause 11.2(26) (Disallowed Cost).

The term 'people' as used by the ECC encompasses the *Contractor's* staff as well as their employees working within the Working Areas.

The cost of people who are directly employed by the *Contractor* in providing the *works* but working outside the Working Areas (e.g. the *Contractor's* staff whose normal place of working is the head office, factory, design office or manufacturing facility) are included in

- SCC cost component 6 (manufacture and fabrication)
- SCC cost component 7 (design) or
- the *fee percentage*.

In general, therefore, the cost of the people who are based at the Site is a cost component in the SCC. Supporting time sheets or daily labour records would show the amount of time that the supervision staff have spent on each project. This is particularly applicable where the *Contractor* is involved in more than one project at the same Site.

There are no direct entries for people required for Contract Data part two since any application for payment would include

- a payroll printout showing the required information
- proof of other payments such as lodging allowances
- other documentary evidence such as invoices (especially for item 14 in the SCC).

The payroll printout would be supported by daily labour records and time sheets so that an audit would reveal a fully traceable cost line. See Section 2.12 and Appendix 3 of Book Two for information about audits.

2.8.3 Cost component 2 – Equipment

Equipment is a defined term in the contract, and comprises items provided by the *Contractor* that are used to Provide the Works, but are not included in the *works* (clause 11.2(9)). The term therefore covers a broad range of items, the two obvious categories being construction plant and temporary works. Examples are excavators, dumpers and generators, scaffolding, and temporary sheet piling and formwork.

Managing Change

In the ECC

- the cost of accommodation is part of Equipment
- the depreciation and maintenance calculations are based on 'open market rates'
- payments for Equipment purchased for work included in this contract are a component of cost.

2.8.3.1 Equipment using the SCC

The SCC includes a number of components for the cost of Equipment, admissible as Defined Cost, such as

- item 21 – hire (externally hired Equipment) or rental of Equipment that is not owned
- item 22 – payments for Equipment that is not listed in the Contract Data but is
  - owned by the *Contractor*
  - purchased by the *Contractor* under a hire purchase or lease agreement or
  - hired by the *Contractor* from the *Contractor's* ultimate holding company or from a company with the same ultimate holding company
- item 23 – payments for Equipment purchased for work included in the contract
- item 24 – payments for special Equipment listed in the Contract Data
- item 25 – consumables (Equipment that is consumed, e.g. fuel)
- item 26 – transporting, erection and dismantling, upgrading and modification
- item 27 – payments for the purchase of materials used to construct or fabricate Equipment.

Item 28 states that the cost of operatives is included in the cost of people, unless included in the hire rates (item 21).

Hire or rent of Equipment not owned by the *Contractor* (item 21)
No Contract Data entries are required for externally hired or rented Equipment.

Any entries in an application for payment should be supported by documentary evidence such as invoices from the hiring or rental company, and plant records or time sheets that show for what activity the item of Equipment was used and for how long.

Payments are made at the hire or rental rate multiplied by the time for which the Equipment is required.

Payments for Equipment not listed in the Contract Data (item 22)
Item 22 states, 'open market rates, multiplied by the time for which the Equipment is required'. There are therefore no Contract Data entries required. Evidence of payment would be required, as well as the time period for which the Equipment was used.

Payments for Equipment purchased for work included in this contract (item 23)
The Contract Data is required to state a time-related charge. The purchase price of the Equipment as well as its value throughout the contract and at Completion is required to be evidenced.

The *Contractor* is required to list in Contract Data part two the Equipment that is purchased for work included in the contract. The time-related charge for the Equipment as well as the time period to which the charge relates should also be included.

The listed items of Equipment purchased for work on the contract, with an on cost charge, are		
Equipment	time-related on cost charge	per time period
....	....	....
....	....	....

Payments for special Equipment listed in the Contract Data (item 24)

The ECC requires the *Contractor* to identify in Contract Data part two any special Equipment that they propose to use in the contract. They are required to provide the time period for which the Equipment is used as well as the rates.

> The rates for special Equipment are
>
> Equipment   rate
>
> ....
>
> ....

The ECC sensibly makes provision for the addition of special items of Equipment to be made to this list given by the *Contractor* at the time of tender in item 24 (if the *Project Manager* agrees, an additional item of special Equipment may be assessed as if it had been listed in the Contract Data). It would seem that the *Contractor* is required to make a request to the *Project Manager*. This is a sensible provision, since contracts are subject to change and there may be a need for special Equipment not envisaged at the outset of the contract.

Consumables (item 25)
No Contract Data entries are required for consumables.

Some items of Equipment may be consumed while the *Contractor* is carrying out the *works*, such as fuels, welding rods and lubricants. The purchase price of these items would be included in an application for payment with the appropriate supporting documentation, such as invoices.

Transport, erection, dismantling, constructing, fabricating or modifying Equipment (item 26)
No Contract Data entries are required for the transportation, erection and upgrading of Equipment.

As long as the costs for the following are not included elsewhere, such as in the hire or rental rates, the *Contractor* may include in an application for payment for the cost of

- transporting Equipment to and from the Working Areas other than for repair or maintenance
- erecting and dismantling Equipment
- constructing, fabricating or modifying Equipment as a result of a compensation event.

Payments for the purchase of materials used to construct or fabricate Equipment (item 27)
This category covers the purchase of materials used to construct or fabricate Equipment, such as the purchase of rolled steel sections, or lifting eyes and chains for modifying the jib of a crane to lift bridge sections.

Cost of operatives (item 28)
Unless included in the hire rates, the cost of operatives is included in the cost of people.

Clearly, the cost of operatives should not appear in the hire rates **and** the cost of people, as these two cost components are mutually exclusive.

2.8.4 Cost component 3 – Plant and Materials

Plant and Materials are items that are intended to be included in the *works* (clause 11.2(14)), such as boilers, turbines, steelwork, pumps, vessels, agitators, cabling, cable trays, concrete and structural steel.

The items of Plant and Materials included in the *Contractor's* Defined Cost would not include items issued to the *Contractor* by the *Client* free of charge.

There are no entries for Plant and Materials required for Contract Data part two, since any application for payment would include invoices and other proof of payment. Note that there are aspects of Disallowed Cost that pertain to Plant and Materials (i.e. there are some items of cost for which the *Contractor* does not get paid).

Since most Plant and Materials tend to be supplied by third parties outsourced by the *Contractor* and are consequently the subject of supply contracts, the Defined Cost of Plant and Materials is relatively easy to identify by reference to the invoices received. Clause 52.1 of the contract makes it clear that all amounts included in the Defined Cost are 'with deductions for all discounts, rebates and taxes which can be recovered'. Disallowed Cost as defined includes the 'cost of Plant and Materials not used to Provide the Works', so it will be necessary to identify any over-ordering by the *Contractor* and to adjust Defined Cost accordingly.

Item 32 makes it clear that cost is credited with payments received for the disposal of Plant and Materials unless the cost is Disallowed.

2.8.4.1 Cost component 4 – Subcontractors

Item 41 introduces into the SCC a cost component for Subcontractors. This covers payments to Subcontractors for work that is subcontracted without taking into account any amounts paid or retained from the Subcontractors by the *Contractor*, which would result in the *Client* paying or retaining the amount twice.

2.8.5 Cost component 5 – charges

This cost component covers a range of items that collectively could be loosely described as site overheads (excluding people) or indirect costs (note that some costs that are traditionally known as site overheads.

2.8.5.1 Charges using the SCC (items 51–54)

The SCC allows for the direct cost for some aspects of charges, such as water, gas and electricity, as well as the rent of premises in the Working Areas. These cost components described within items 51, 52, 53 and 54 of the SCC would, for the most part, be supportable by documentary evidence, and therefore there are no Contract Data part two entries required for these cost components.

2.8.6 Cost component 6 – manufacture and fabrication

The cost components for manufacture and fabrication outside the Working Areas include the cost of manufacture or fabrication of Plant and Materials that are

- wholly or partly designed specifically for the *works* and
- manufactured or fabricated outside the Working Areas.

This cost component excludes the costs of manufacture or fabrication of Plant and Materials that are 'off the shelf' (which would appear under cost component 3 of the SCC).

The *Contractor* tenders in Contract Data part two an hourly rate for the categories of their own employees who would work in a workshop or factory outside the Working Areas.

The rates for Defined Cost of manufacture and fabrication outside the Working Areas by the *Contractor* are

category of person   rate

....

....

Schedule of Cost Components and Short Schedule of Cost Components

The use of cost-reimbursable contracts is not recommended where manufacture or fabrication outside the Working Areas forms a major part of a contract because of the difficulty this presents in the control and identification of Defined Cost. The same could be said to apply where design outside the Working Areas forms a major part of a contract. For this reason, the ECC takes a very cautious approach, relying on the tendered hourly rates inserted by the *Contractor* in Contract Data part two as the basis for calculating the Defined Cost of these activities. It only requires, therefore, for the *Project Manager* to satisfy themselves as to the time spent by the *Contractor* on these activities.

A possible solution in this situation would be to consider having an Option A fixed-price contract for the manufacture and supply of, say, the pipework. However, this would still not remove the problems associated if compensation events arise.

2.8.7 Cost component 7 – design

Cost component 7 covers design of the *works* and Equipment done outside the Working Areas.

The *Contractor* tenders in Contract Data part two the category of their own employees who will work on the design, as well as the categories of employees who would travel to and from the Working Areas for the purposes of design.

> The rates for Defined Cost of design outside the Working Areas are
>
> category of person   rate
>
> ....
>
> ....
>
> The categories of design people whose travelling expenses to and from the Working Areas are included as a cost of design of the *works* and Equipment done outside of the Working Areas are
>
> ....
>
> ....

2.8.8 Cost component 8 – insurance

Cost component 8 of the SCC provides for the following to be deducted from Defined Cost:

- the cost of events for which this contract requires the *Contractor* to insure
- other costs paid to the *Contractor* by insurers.

The first avoids the *Client* having to pay for costs that the *Contractor* should have insured against. If the *Contractor* does not insure as required by the contract, then such costs are at their own risk. An example of the first category is loss of or damage to Equipment, which clause 83.2 requires the *Contractor* to insure against. If a piece of Equipment owned and being used by the *Contractor* to Provide the Works catches fire and is destroyed, then its replacement cost is deducted from the Defined Cost. This avoids the *Client* having to pay for costs that the *Contractor* should have insured against. If the *Contractor* does not insure as required by the contract, then such costs are at their own risk. In practice, such an eventuality should never arise if the *Project Manager* requests evidence from the *Contractor* that the required insurances are in force (refer to clause 84.1).

The second deduction ensures that the *Contractor* does not receive double payment as a result of, for example, insurance that they have voluntarily taken out or from insuring for a greater cover than required by the contract. The practical complication with this is that, unless the *Contractor* volunteers the information, the *Client* will not know the scope of any difference in cover insurance that the *Contractor* has effected.

There are no Contract Data entries required for this cost component heading.

2.9. The components of cost included under the SSCC

Used with main Options A and B.

2.9.1 Introduction

The opening statement at the beginning of the SSCC states:

'This schedule is part of these *conditions of contract* only when Option A or B is used. An amount is included

- only in one cost component and
- only if it is incurred in order to Provide the Works.'

This opening statement to the SSCC makes it clear that it is part of the *conditions of contract* **only** when main Option A or B and is used for the assessment of compensation events, unless it is agreed to use rates and lump sums (clause 63.2).

As a general policy, the SSCC only provides for the direct reimbursement of those cost components that are readily identifiable. The principal way this is achieved is through the concept of Working Areas: areas of land comprising the Site that are made available by the *Client*, together with any additional areas proposed by the *Contractor* and accepted by the *Project Manager* as necessary to Provide the Works.

Generally, only the cost of People and Equipment doing work in these Working Areas is reimbursed as Defined Cost. This recognises the difficulty and thus the higher risk to the *Client* in identifying and controlling costs that are incurred away from the Site (i.e. outside the Working Areas).

The exception to this general principle is those costs incurred by the *Contractor* associated with the design, manufacture and fabrication outside of the Working Areas that are dealt with separately in the SSCC (in cost components 6 and 7).

Recognising that there could be some overlap between various cost components, the SSCC sensibly provides the reminder that 'amounts are included only in one cost component'. The SSCC provides for both direct reimbursement in GBP (or other currency) and also for indirect reimbursement, by the use of predetermined percentages to cover a range of cost components.

To make the SSCC fully effective requires the insertion of certain data by the *Contractor* in Contract Data part two. It is important that all data required by Contract Data part two for the SSCC is included by the *Contractor* **before submitting their tender**. For main Options A and B, the data is used in the assessment of compensation events.

There are eight headings for the components of cost included in the Shorter SCC:

1 'People'
2 'Equipment'
3 'Plant and Materials'
4 'Subcontractors'
5 'Charges'
6 'Manufacture and fabrication (outside the Working Areas)'
7 'Design (outside the Working Areas)'
8 'Insurance'.

Under each of these headings are described all the components of cost for which the *Contractor* will be reimbursed during the period of the contract and which will also be taken into consideration when evaluating the financial effects of compensation events. Any other components of cost not identified in the schedule should be covered, and are deemed anyway to be covered by inclusion in the *fee percentage*.

2.9.2 Cost component 1 – people

The SSCC identifies three components of cost for people for which the *Contractor* has to provide evidence of amounts paid by them, including those for meeting the requirement of the law and for pension provisions:

- People directly employed by the *Contractor* and whose normal place of working is within the Working Areas; that is, direct employees of the *Contractor* whose normal place of working is within the Working Areas (e.g. tradesmen or the site agent).
- People directly employed by the *Contractor* and whose normal place of working is **not** within the Working Areas but who are working in the Working Areas proportionate to the time they spend working in the Working Areas; that is, direct employees of the *Contractor* whose normal place of working is not within the Working Areas but who are working in the Working Areas (e.g. a specialised tradesman who is being used for a compensation event).
- People directly employed by the *Contractor* (and who are not Subcontractors as defined) but are paid for by them according to the time worked while they are within the Working Areas.

The inclusion of the last category reflects the more normal practice of paying such people on agreed hourly or daily rates.

The People Rates are identified in Contract Data part two as follows:

The people rates are		
Category of person	unit	rate
....	....	....
....	....	....

The term 'people' as used by the ECC encompasses the *Contractor's* staff as well as their employees working within the Working Areas.

The cost of people who are directly employed by the *Contractor* in providing the *works* but working outside the Working Areas, such as the *Contractor's* staff whose normal place of working is the head office, factory, design office or manufacturing facility, are included in either cost component 6 or 7.

2.9.3 Cost component 2 – equipment

2.9.3.1 Equipment using the SSCC

The SSCC components for the cost of Equipment are as follows:

- item 21 – Equipment included in a published list (e.g. the Civil Engineering Contractors Association (CECA) Daywork Schedule); Equipment includes the cost of the *Contractor's* accommodation
- item 22 – Equipment not included in a published list
- item 23 – the time required
- item 24 – transporting, erection and dismantling, upgrading and modification
- item 25 – consumables
- item 26 – the cost of operatives
- item 27 – Equipment that is neither in the published list stated in the Contract Data nor listed in the Contract Data.

Equipment in a published list (item 21)

Contract Data part two requires entries for items included in a published list of Equipment:

The published list of Equipment is the edition current at the Contract Date of the list published by ....
The percentage for adjustment for Equipment in the published list is ....% (state plus or minus)

Equipment not in a published list (item 22)
Contract Data part two requires a rate as well as a description for items not included in a published list of Equipment:

```
The rates of other Equipment are
Equipment                                              rate
..............................................        ....................
..............................................        ....................
```

The time required (item 23)
This item outlines how the time required for an item of Equipment is to be expressed:

> 'The time required is expressed in hours, days, weeks or months consistently with the list of items of Equipment stated in the Contract Data or with the published list stated in the Contract Data.'

Transporting, erection and dismantling, constructing, fabricating or modifying Equipment (item 24)
No Contract Data entries are required for the transportation, erection and upgrading of Equipment.

As long as the costs for the following are not included elsewhere, such as in the hire rates or the depreciation and maintenance charge, the *Contractor* may include in an application payment for the cost of

- transporting Equipment to and from the Working Areas other than for repair and maintenance
- erecting and dismantling Equipment
- constructing, fabricating or modifying Equipment as a result of a compensation event.

Consumables (item 25)
No Contract Data entries are required for consumables unless the purchase price of the Equipment that is consumed is in a published list stated in the Contract Data.

Cost of operatives (item 26)
Unless included in the hire rates, the cost of operatives is included in the cost of people. Clearly, the cost of operatives should not appear in the hire rates **and** the cost of people, as these two cost components are mutually exclusive.

Equipment not in the published listed or listed in the Contract Data (item 27)
Where Equipment is neither in a published list stated in the Contract Data nor listed in the Contract Data, then this Equipment is assessed 'at competitively tendered or open market rates'. The onus will be on the *Contractor* to demonstrate that the rates put forward fulfil these criteria.

2.9.4 Cost component 3 – Plant and Materials

Plant and Materials are items that are intended to be included in the *works* (clause 11.2(14)): for example, boilers, turbines, steelwork, pumps, vessels, agitators, cabling, cable trays, concrete and structural steel.

The items of Plant and Materials included in the *Contractor's* Defined Cost would not include items issued to the *Contractor* by the *Client* free of charge.

The cost components for Plant and Materials as described in the SSCC comprise payments for

- purchasing Plant and Materials
- delivery to and removal from Working Areas
- providing and removing packaging
- samples and tests.

There are no entries for the Plant and Materials required for Contract Data part two, since any application for payment would include invoices and other proof of payment. Note that there are aspects of Disallowed Cost that pertain to Plant and Materials (i.e. there are some items of cost for which the *Contractor* does not get paid).

Since most of Plant and Materials tend to be supplied by third parties outsourced by the *Contractor* and consequently the subject of supply contracts, the Defined Cost of Plant and Materials is relatively easy to identify by reference to the invoices received. Clause 52.1 of the contract makes it clear that all amounts included in Defined Cost are 'with deductions for all discounts, rebates and taxes, which can be recovered'. Disallowed Cost as defined includes the 'cost of Plant and Materials not used to Provide the Works', so it will be necessary to identify any over-ordering by the *Contractor* and to adjust Defined Cost accordingly.

Item 32 makes it clear that cost is credited with payments received for the disposal of Plant and Materials unless the cost is disallowed.

2.9.5 Cost component 4 – Subcontractors

Item 41 introduces into the SCC a cost component for Subcontractors, covering payments to Subcontractors for work that is subcontracted.

2.9.6 Cost component 5 – charges

The SSCC caters for the direct cost for some aspects of charges:

'51 payments for the provision of and use in the Working Areas of
- water
- gas
- electricity
- telephone and
- internet

52 Payments to public authorities and other properly constituted authorities of charges which they are authorised to make as part of the *works*

53 Payments for
  (a) cancellation charges arising from compensation events
  (b) buying or leasing land or buildings within the Working Area
  (c) compensation for loss of crops or buildings
  (d) royalties
  (e) inspection certificates
  (f) charges for access to the Working Areas
  (g) facilities for visits to the Working Areas by Others
  (h) consumables and equipment provided by the *Contractor* for the *Project Manager's* and *Supervisor's* offices

54 Payments made and received by the *Contractor* for the removal from Site and disposal or sale of materials from excavation and demolition.'

2.9.7 Cost component 6 – manufacture and fabrication

The components of cost for manufacture and fabrication are exactly the same as in the SCC.

2.9.8 Cost component 7 – design

The components of cost for design are exactly the same as in the SCC.

2.9.9 Cost component 8 – insurance

The components of cost for insurance are exactly the same as in the SCC.

2.10. Contract Data part two

Note that the data for the SCC and SSCC shown in the individual component cost sections above do not appear in the same order in Contract Data part two.

2.11. Putting it all together for payment – Option C

Once the cost of the each of the component headings can be ascertained, it is a matter of adding up the costs for each cost component. A suggested format is provided in Table 2.3; however, each *Contractor* should use the calculation that most suits them.

The final calculation for presentation in an application for the amount due for payment in the ECC is shown in Table 2.4.

Table 2.3 Putting the SCC together

Cost component		Comment	£ total
1	**People**		
11 12 13	Components for items 11, 12 and 13 People who are directly employed by the *Contractor* and whose normal place of working is within the Working Areas	Payroll sheets for the period concerned	£
11 12 13	Components for items 11, 12 and 13 People who are directly employed by the *Contractor* and whose normal place of working is not within the Working Areas but who are working within the Working Areas	Payroll sheets for the period concerned	£
14	The following components of the cost of people who are not directly employed by the *Contractor* but are paid by the *Contractor* according to the time worked while they are within the Working Areas Amounts paid by the *Contractor*	Proof of payment of the amounts made	£
2a	**Equipment for the SCC** **Note**: In the ECC the definition of Equipment includes accommodation		
21	Payments for the hire or rent of Equipment not owned by ■ the *Contractor* ■ the *Contractor's* ultimate holding company or from a company within the same ultimate company	Invoices and proof of payment – the hire rate or rental rate multiplied by the time for which the Equipment is required	£
22	Payments for Equipment that is not listed in the Contract Data but is ■ owned by the *Contractor* ■ purchased by the *Contractor* under a hire purchase or lease agreement or ■ hired by the *Contractor* from the *Contractor's* ultimate holding company or from a company with the same ultimate holding company	Invoices and proof of payment – at open market rates multiplied by the time for which the Equipment is required	£
23	Payments for Equipment purchased for work included in the contract listed with a time- related on-cost charge, in the Contract Data, of ■ the change in value over the period for which the Equipment is required ■ the time-related on-cost charge stated in the Contract Data for the period for which the Equipment is required	Proof of purchase Time-related on-cost from Contract Data Demonstration of the change in value over period	£

Cost component		Comment	£ total
24	Payments for special Equipment listed in the Contract Data	Rates in the Contract Data multiplied by the time for which the Equipment is required	
25	Payments for the purchase price of Equipment that is consumed	Invoice and proof of payment	£
26	Unless included in the hire or rental rates, payments for ■ transporting Equipment to and from the Working Areas other than for repair and maintenance ■ erecting and dismantling Equipment ■ constructing, fabricating or modifying Equipment as a result of a compensation event	Documentary proof of payment	£
27	Payments for purchase of materials used to construct or fabricate Equipment	Documentary proof of payment	£
28	Unless included in the hire rates, the cost of operatives is included in the cost of people	Not calculated	0
2b	**Equipment for the Short SCC**		
21	Amounts in the published list stated in Contract Data part two	Rates in the published list in the Contract Data multiplied by the percentage adjustment in Contract Data part two, multiplied by the time for which the Equipment is required	£
22	Amounts not in the published list stated in Contract Data part two	Rates in Contract Data part two multiplied by the time for which the Equipment is required	£
23	The time required is expressed in hours, days, weeks or months consistent with the list of items of Equipment in the Contract Data or with the published list stated in the Contract Data	Statement of how to calculate time for the Equipment	0
24	Unless included in the published list, payments for ■ transporting Equipment to and from the Working Areas other than for repair and maintenance ■ erecting and dismantling Equipment ■ constructing, fabricating or modifying Equipment as a result of a compensation event	Invoice and proof of payment	£
25	The purchase price of Equipment that is consumed, unless in the published list or the rate includes the purchase price	Documentary proof of payment	£
26	The cost of operatives is always included either in cost component 1 or 21 or 22	Statement of where to include people costs	0
27	Amounts for Equipment which is neither in the published list stated in the Contract Data nor listed in the Contract Data, at competitively tendered or open-market rates, multiplied by the time for which the Equipment is required	Invoice and proof of payment	£
3	**Plant and Materials**		
31	Payments for ■ purchasing Plant and Materials ■ delivery to and from the Working Areas ■ providing and removing packaging ■ samples and tests	Invoice and documentary proof of payment	£
32	Payments received for the disposal of Plant and Materials	Documentary proof of payment	−£
4	**Subcontractors**		
41	Payments made to Subcontractors	Documentary proof of payment	£

	Cost component	Comment	£ total
5a	**Charges for the SCC**		
51	Payments for provision and use in the Working Areas of ■ water ■ gas ■ electricity ■ telephone ■ internet	Invoice and proof of payment	£
52	Payments to public authorities and other properly constituted authorities of charges which they are authorised to make in respect of the *works*	Invoice and proof of payment	£
53	Payments for (a) cancellation charges arising from a compensation event (b) buying or leasing land or buildings within the Working Area (c) compensation for loss of crops or buildings (d) royalties (e) inspection certificates (f) charges for access to the Working Areas (g) facilities for visits to the Working Areas by Others (h) consumables and equipment provided by the *Contractor* for the *Project Manager's* and *Supervisor's* offices	Invoice or other document and proof of payment	£
54	Payments made and received by the *Contractor* for the removal from Site and disposal of materials from excavations and demolitions	Invoice and proof of payment	£
6a	**Manufacture and fabrication outside the Working Areas for the SCC**		
61 62	Cost of manufacture or fabrication of Plant and Materials that are wholly or specifically designed for the *works* and that are manufactured or fabricated outside the Working Areas	(Hourly rates in Contract Data part two multiplied by hours worked)	£
7	**Design outside the Working Areas**		
71 72	Cost of design of the *works* and Equipment done outside the Working Areas	Hourly rates in Contract Data part two multiplied by the hours worked	£
73	Cost of travel to and from the Working Areas for the categories of employees listed in Contract Data part two	Invoice/tickets or business mileage plus proof of payment	
8	**Insurance**		
	The cost of events for which the contract requires the *Contractor* to insure	Documentary proof of payment	−£
	Other costs paid to the *Contractor* by insurers	Documentary proof of payment	−£
		Total of the SCC	£

Table 2.4 Presentation for the ECC application for payment

	Direct work	
1	Total of the cost components in the SCC	£
2	Less Disallowed Cost	−£
3	Total Defined Cost	£
4	Less Disallowed Cost	−£
5	*Fee percentage*	%
6	**Other amounts**	
7	**plus** other amounts to be paid to the *Contractor* (e.g. interest on late payments)	£
8	**less** other amounts to be paid or retained from the *Contractor* (e.g. retention and other amounts)	−£
9	**less** previous amounts due for payment	−£
10	**Total amount due for this application**	£

2.12. Inspecting accounts and records

2.12.1 Compensation events

Because the SCC and SSCC are focused on the pre-assessment of change maybe many months ahead of the actual work being undertaken, a quotation for a compensation event may therefore not always include supporting documentation such as quotations/invoices for Plant and Materials or wage slips.

If your project is required to be audited, either by an internal or external body, then you should ensure that the departments/bodies involved realise this at the very outset of the project. It is also suggested that you consult with them at the earliest moment, so that any requirements or implications that may affect the way you operate the contract can be considered and, where appropriate, included in the Scope.

2.12.2 Options C, D and E

It may not be possible for the *Project Manager* to examine fully all the information in a *Contractor's* application for payment within the 1 week required to issue a payment certificate. The *Project Manager* may only have sufficient time to perform a spot check of the supporting documentation, and would have to rely on later audits of the *Contractor's* books to check the Price for Work Done to Date and possibly the final total of the Prices.

It is recommended that the contract requires the *Contractor* to set out their application for payment in a certain way to assist the *Project Manager* in their assessment of the amount due. Clause 50.2 requires the *Contractor* to submit an application for payment. Assessment of the amount due would be virtually impossible if the *Contractor* does not provide all the information in an amount due prior to the *assessment date*. It would be to the *Contractor's* benefit as well to ensure that all their costs were visible and easily traceable through the supporting documentation, so that disputes over valuations are kept to a minimum.

Clause 50.4 states that

> 'If the *Contractor* does not submit an application for payment before the assessment date, the amount due at the assessment date is the lesser of
>
> - the amount the *Project Manager* assesses as due at the assessment date, assessed as though the *Contractor* had submitted an application before the assessment date, and
> - the amount due at the previous assessment date.'

This is a powerful incentive for the *Contractor* to submit an application for payment.

For example, in the people section of the SCC (cost component 1), where the *Contractor* lists their on-site labour, including supervision, they could list the people by name and job description, the job performed and the hours spent on that job. This would generally be a copy of the daily labour record in a more legible and visible format. This would be particularly important where Disallowed Costs are incurred, such as the correction of Defects, so that the *Project Manager* is able to see where the *Contractor* has deducted the time spent on the elements making up Disallowed Costs. This also underlines the importance of having the *Project Manager* on Site, in order to be aware of the performance on Site. There is, of course, an element of mutual trust and co-operation running through the contract, particularly for the application for payment, although the *Project Manager* has the ability to correct payment certificates later (clause 51.3).

Audits can pick up small things that the *Project Manager* might miss during their assessment, and it could also pick up irregularities that are not noticeable in each individual application for payment but are manifest over a period of time. At the very least, an audit could reveal whether invoices claimed for under the contract have actually been paid later on. It is recommended that at least one audit is conducted for Option C, D and E contracts, and many more for longer projects.

An example of an audit plan is included in Appendix 3 of Book Two.

2.13. Use of the SSCC

The SSCC is **only** used for main Options A and B. The SSCC is simpler than the SCC used for main Options C, D and E:

- cost component 1 – people uses People Rates that are identified by the *Contractor* in Contract Data part two
- cost component 2 – a published list is used for Equipment costs
- cost component 6 – amounts paid by the *Contractor*.

A comparison between the SSC and the SSCC is given in Figure 2.2.

2.14. Practical issues
2.14.1 Working on multiple projects on the same site

The ECC SCC does not necessarily cater very well for contracts where there are many contractors on the same site. Because many of the percentages are based on projected turnover and the costs of the site, the *Contractor* may only be able to be realistic with percentages if the *Client* has advised them of the projects that the *Contractor* will be performing over a period.

2.14.2 Example of the principles of the assessment of change

A new retaining wall is to be constructed as shown in Figure 2.3. This is a retaining wall on a *Client*-designed project. It is realised, however, that the length of the retaining wall needs to be increased from 10 to 20 m. The *Project Manager* acknowledges that this is a change to the Scope, and raises an instruction and compensation event notification in which they instruct the *Contractor* to submit a quotation.

Table 2.5 shows three possible scenarios: A, B and C.

The main Option for the contract is Option A (the same principles would work for Option B as well). The correct tendered price for the work is £2000 – scenario B. In scenario A, the *Contractor* has underpriced the true value of the original work in their tender, and has inserted £1000. In scenario C a high price has been inserted of £3000. It should be noted that the prices against individual items should be set against the context of the pricing for the whole contract. The example of low, correct and high is given here for illustrative purposes on the principles of the SCC.

For the purposes of assessing the changes to the Prices from a compensation event to change the length of the retaining wall, the original tendered prices of £1000 (low) in scenario A, £2000 (correct) in scenario B and £3000 (high) in scenario C are not used. Instead, the original work and the revised work are priced using the SCC.

In this example, as shown in Table 2.5, the assessment of the original work using the SCC will give a true assessment of £2000 in all three scenarios. This figure is then compared with the

Figure 2.2 The components of Defined Cost for the SSC and SSCC

SCC	SSCC

1. People
Cost components 11, 12 and 13:
• directly employed people
   – whose normal place of work is the Working Area
   – whose normal place of work is not the Working Area
11. Wages and salaries
12. Payments related to work on the contract and made to people for (a) to (f)
13. Payments made in relation to people in accordance with their employment contract for (a) to (o)
14. Non-direct employed people. Amounts paid by the *Contractor*
	1. People
	Cost component 11:
	• directly employed people
	   – whose normal place of work is the Working Area
	   – whose normal place of work is not the Working Area
	• not directly employed people but who are paid while they are in the Working Area
	Amounts calculated by multiplying each of the People Rates by the time appropriate to the rate spent in the Working Areas.

2. Charges 21. Hire or rent of Equipment not owned by *Contractor* 22. Equipment not listed in the Contract Data but owned by the *Contractor* 23. Equipment purchased for the contract 24. Special Equipment 25. Consumables 26. Transporting, etc. 27. Materials to fabricate Equipment 28. Unless included in the published list, cost of operatives is included in the cost of people	**2. Charges** 21. Published list (e.g. CECA) 22. List stated in Contract Data 23. Time expressed in hours, days, weeks or months 24. Transporting, etc. 25. Consumables 26. Unless included in the published list, cost of operatives is included in the cost of people. 27. Open market rates for Equipment not in the published list or the list stated in the Contract Data
3. Plant and Materials (At cost)	**3. Plant and Materials** (At cost)
4. Subcontractors Payments to Subcontractors	**4. Subcontractors** Payments to Subcontractors
5. Equipment The following charges are identified separately: 51. Water, gas, electricity, telephone and internet 52. Public authorities 53. Cancellation charges, etc., (a) to (h) 54. Removal from the Site and disposal or sale of materials from excavation and demolition	**5. Equipment** The following charges are identified separately: 51. Water, gas, electricity, telephone and internet 52. Public authorities 53. Cancellation charges, etc., (a) to (h) 54. Removal from the Site and disposal or sale of materials from excavation and demolition
6. Manufacture and fabrication 61. Total hours worked by employees by hourly rates stated in the Contract Data	**6. Manufacture and fabrication** 61. The same as the SCC
7. Design outside of the Working Areas The following components of cost of design of the works and Equipment done outside the Working Areas 71. Amounts calculated by multiplying each of the rates for people in the Contract Data by the total time appropriate to that rate spent on design of the works and Equipment outside the Working Areas in the Contract Data 72. Cost of travel to and from the Working Areas for the categories of design people listed in the Contract Data	**7. Design** 71. The same as the SCC
8. Insurance Deducted from the cost: • costs against which the contract required the Contractor to insure • Others costs paid to the Contractor by insurers	**8. Insurance** Deducted from the cost: • cost of events for which the contract requires the *Contractor* to insure • Others costs paid to the *Contractor* by insurers
Compensation events only *Contractor*'s risks (clause 63.8)	**Compensation events only** *Contractor*'s risks (clause 63.8)
The Fee *Fee percentage* as stated in Contract Data part two	**The Fee** *Fee percentage* as stated in Contract Data part two

assessment of the revised work to the retaining wall of £4000 in each scenario; taking one from the other gives a total change to the Prices of £2000 in all three scenarios.

In each scenario the outcome is the same when using the SCC to assess the changes to the Prices by omitting the original work and adding in the revised work, so that the outcome is the same in each instance. This therefore removes the arguments about the use and applicability of rates, prices and lump sums submitted at the time of tender.

Figure 2.3 Cross-section through a reinforced concrete retaining wall

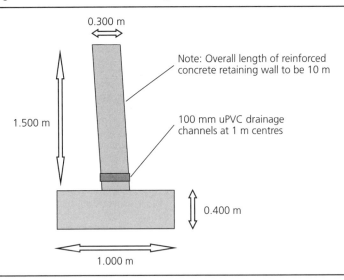

2.14.3 Omissions

If a *Client* omits work from a contract, the omission will be assessed using the SCC or shorter (depending on the main Option) rather than simply omitting the relevant sums in the Activity Schedule or the *bill of quantities*. The build-up for the omission in the form of a quotation will include the tendered *direct fee percentage*.

This raises the issue of loss of profit on omitted work for the *Contractor*. The ECC is simply silent on this matter. From a practical point of view small omissions are of little consequence, unless they build up to such an extent that they become a large change or omission to the scope of the *works*. Larger omissions, which affect the overall scope of the *works*, are a different matter.

The ECC is based on the concept of the *Client* planning their *works* well. Nevertheless, circumstances do occur when, no matter how well planned a project, the project is overtaken by events. Let us consider an example where the *Client* owns a complex of buildings on one Site.

The project involves the construction of a new five-storey office block, which is to be linked to an existing office block by a subway under the site link road to an existing basement entrance in the existing office (Figure 2.4). This connection to the existing office block has been identified as a separate activity on the Activity Schedule in an Option A contract.

Table 2.5 SCC low-, correct and high-price scenarios

	Scenario A: low price	Scenario B: correct price	Scenario C: high price
Tendered prices			
Item Original retaining wall (This could be a price for an activity in an *activity schedule* or a *bills of quantities* item)	£1000	£2000	£3000
Total	**£1000**	**£2000**	**£3000**
Assessment of change using the SCC			
Assessment of the original work	£2000	£2000	£2000
Assessment of the revised work	£4000	£4000	£4000
Total of the changes to the Prices	**£2000**	**£2000**	**£2000**

Figure 2.4 Office block scheme

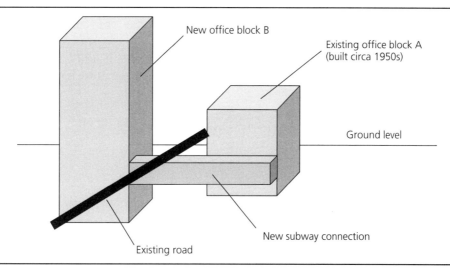

Twelve months into the project the *Client's* facilities management team has identified that, due to rapid growth, the existing office facilities need to be increased to cope with this growth. It has also been identified that the existing office facilities are now below modern standards, and the team has recommended that the existing office block is demolished and a new office built in its place.

The *Client* has also decided that it would be more appropriate and less disruptive to their operations on Site if the subway connection were repackaged into the new project. The work on the subway is still some 6 months away.

The *Project Manager* issues an instruction and compensation event notification omitting the new subway. The *Contractor* sends a quotation, which equals the value of £250 000 shown on their tendered Activity Schedule for the project.

The *Project Manager* does not accept the quotation, and instructs the *Contractor* to provide a revised quotation. The reason given is that the quotation is not in accordance with clause 63.1, namely that that quotation has not been assessed on

- the actual Defined Cost of the work already done by the dividing date
- the forecast Defined Cost of the work yet to be done by the dividing date
- the resulting Fee.

The *Contractor* resubmits their quotation on this basis and in accordance with the contract. The total omission is £243 348. This is the amount by which the Prices would be reduced according to the compensation event, rather than the £250 000 originally included in the Activity Schedule.

2.14.4 The *Project Manager's* assessment

The contract contains within it the facility for the *Project Manager* to make their own assessment. If we imagine in the office block example that the *Contractor* refuses to provide a revised quotation, then the *Project Manager* can make their own assessment (see also Section 1.9.1 of Chapter 1 in this Book). The *Project Manager's* assessment comes to £298 000. This means that the amount by which the Prices would be reduced according to the compensation event would be £298 000, rather than the £250 000 originally included in the Activity Schedule (always remembering that the price inserted against an activity may not be reflective of its true price). Since the reduction is larger than the originally included figure, it could be assumed that the *Contractor's* profit will be affected by this compensation event.

The figure of £298 000 is substantially higher than the one that the *Contractor* would have come to if they had done the quotation themselves. It should be noted that the *Project Manager* does not have to instruct the *Contractor* to resubmit their quotation – they could go straight to a

Managing Change

Project Manager's assessment. It is unwise for the *Contractor* to get to such a point, since the *Project Manager* will not have the same level of information in regard to the compensation event as the *Contractor* themselves.

This being the case, and as long as they demonstrate that they have built up the quotation as required by the contract using reasonable skill and care, there is no reason why their assessment should not be acceptable.

Areas that will have a major influence on the assessment are

- programme durations
- output levels
- the critical path
- resource levelling, either on the whole or parts of the *works* for both People and Equipment.

> If a *Contractor* submits a quotation that is not in accordance with the requirements of the contract, the *Project Manager* can make their own assessment. It is not a requirement of the contract that the *Project Manager* gives the *Contractor* an opportunity to resubmit a quotation before they make their own assessment.

2.14.5 Numerous small compensation events

The ECC assumes that every compensation event is assessed individually. However, in the hurly-burly of everyday projects this is rarely the case. Sometimes once one thing goes wrong or a dimension is changed, it triggers a whole sequence of events, albeit each one being very minor.

In such an instance it may be appropriate for a number of these small compensation events on related items to be grouped and assessed together (Table 2.6). In this way it is possible to identify any possible knock-on effects that may not be evident from each single small compensation event. It will also facilitate picking up what has been traditionally called the 'disruption' element of change.

2.14.6 Issue of Site Information drawings

The *Project Manager* may issue an instruction containing some revised Site Information drawings labelled 'For information purposes only'.

The issue of Site Information drawings in itself is not a compensation event unless the issue of that information requires a change to the Scope, in which case it will be a compensation event.

Project Managers should be mindful to issue only relevant Site Information drawings and not just issue a blanket set of Site Information drawings. Hopefully, the *Project Manager* will have reviewed the drawings to see if they have or are likely to have any implications on the Scope.

2.14.7 Occasions when Defined Cost is not used

There are certain occasions when the contract does not call for the use of Defined Cost as defined in the SCC.

Table 2.6 Compensation events grouped together for assessment

Lift shaft A				
PMI No.	CE No.	Description	Effect	
			Time: days	Price: £
1	3	Additional reinforcement to waling beam A	0.25	1000.00
4	7	Missing rebar to waling beam B	0.50	500.00
8	9	Cut rebar on site to revised dimensions	0.75	500.00
		Total	**1.50**	**2000.00**

Schedule of Cost Components and Short Schedule of Cost Components

2.14.7.1 Uncorrected Defects

If a notified Defect is not corrected, the *Project Manager* assesses the **cost** of having the Defect corrected by other people. The **cost** will be whatever that is as assessed by the *Project Manager*.

> The *Project Manager* notifies the *Contractor* that they have not corrected the defective plasterwork in the entrance area of the new hotel within the *defect correction period*. The *Project Manager* therefore assesses the **cost** of having the Defect corrected by other people, and advises the *Contractor* of this fact.

2.14.7.2 Access to the Site

Any cost incurred by the *Client* as a result of the *Contractor* not providing the facilities and services they are to provide is assessed by the *Project Manager* and paid by the *Contractor*.

> In the section for facilities and services to be provided by the *Contractor*, the Scope requires the *Contractor* to provide a cleaner for the site accommodation. Four weeks after the start of the contract, no cleaner has appeared, even after repeated requests by the *Project Manager*. The *Project Manager* advises the *Contractor* that they have hired a cleaner for the duration of the contract or until such time as they provide the cleaner as required by the Scope. The *Project Manager* has assessed this cost to be £20 per day, and they advise the *Contractor* that they will be required to pay this cost as detailed in clause 25.2.

2.14.7.3 Tests and inspections

The cost incurred by the *Client* in repeating tests after a Defect is found is assessed by the *Project Manager* and paid by the *Contractor*.

> The *Contractor* offers up some completed watermain pipework as being completed and free from Defects. The Scope requires the *Client* to carry out water tests on the pipework. The pipework fails the test. A week later, the pipework, after being corrected, is retested and passes the *Client's* test. The *Project Manager* notifies the *Contractor* that they have assessed the costs incurred by the *Client* in redoing the test, and advises that £896.00 is to be paid by the *Contractor*.

2.14.7.4 Acceleration

If the *Project Manager* has instructed the *Contractor* to submit a quotation for acceleration and the *Contractor* chooses to do so, the quotation they submit is not required to be based on Defined Cost.

> **Question**
> A *Project Manager* becomes aware after the contract is let that the data for the SCC and SSCC in Contract Data part two have not been completed or that the *Contractor* is struggling to complete the information. What should they do?
>
> **Answer**
> Strictly this is the *Contractor's* problem; however, their failing to understand the SCC may cause problems later in the project. So it may be in the interests of all involved to ensure that they have understood the requirements of the contract.

2.14.7.5 Quotations manual

The cost of people involved in a compensation event is based on the components of cost for people listed in the SCC. If there are few compensation events, then this process is straightforward. However, if you have many compensation events the calculation of people costs for each and every compensation event may be very time consuming. To overcome this, it may be more practical to establish a quotations manual in which the initial People Rates (and other components of cost) have been agreed and calculated using the SCC at the outset of the contract. This has the advantages of

- establishing agreed rates at the outset of the contract
- providing consistency on multiple Sites/administration sites
- speeding up the assessment of compensation events.

2.14.7.6 Disallowed Cost

Finally, recognising the *Client's* potential vulnerability the SSCC includes for People Rates under cost-based contracts, the ECC includes the concept of **Disallowed Cost**, a full and lengthy definition of which is provided in main Option clauses C11.2(26), D11.2(26), E11.2(26) and F11.2(27). Generically, the definition is a list of things for which the *Contractor* will not be reimbursed; that is, any costs incurred against the headings identified will be deducted from Defined Cost. Most of the things included could be said to derive from some 'shortcoming' of the *Contractor* or failure to conduct their operations to acceptable standards. This immediately introduces an element of discretion, which falls to be exercised by the *Project Manager*, a subject that has recently turned the spotlight on the *Project Manager's* implied duty to act impartially and in good faith. Examples of Disallowed Cost include

- cost that the *Project Manager* decides is not justified by the *Contractor's* accounts and records
- the cost of correcting Defects after Completion
- Plant and Materials not used to Provide the Works (after allowing for reasonable wastage)
- resources not used to Provide the Works (after allowing for reasonable availability and utilisation) or not taken away from the Working Areas when the *Project Manager* requested.

Although payments to Subcontractors for work that is subcontracted constitute Defined Cost, lest *Contractors* run away with the idea that their administration of subcontracts can be to some lesser standard, Disallowed Cost includes the following Subcontractor-specific checks:

- cost which should not have been paid to a Subcontractor in accordance with their Subcontract
- cost that the *Project Manager* decides results from paying a Subcontractor more for a compensation event than is included in the accepted quotation or assessment for the compensation event.

So, given that the *Contractor* is clearly at risk that some of their Defined Cost will not be reimbursed, how should they cover themselves against this eventuality, given, for example, that some Disallowed Cost is almost inevitable (e.g. the cost of correcting Defects after Completion)? The answer while simple is not always obvious. Any Defined Cost that the *Contractor* anticipates incurring but which may be the subject of a Disallowed Cost deduction has to be recovered through the Fee, and consequently the *fee percentage* will need to include an allowance for protecting against this risk. Clearly, some *Clients*, depending on their choice of *Project Manager*, will be regarded by *Contractors* as more 'risky' than Others, and this may be reflected in the tendered *fee percentage*.

2.15. Preliminaries and people costs
2.15.1 Introduction

Appendix 3 in this book provides practical examples of how people costs are built up and how the traditional calculation compares to the SCC and SSCC, and how this relates to the *fee percentage*.

Appendix 4 gives a comparison of a traditional preliminaries build-up for the SCC and SSCC.

Both of these appendices serve to clarify the interrelationship between traditional practice and the ECC.

Managing Change
ISBN 978-0-7277-6188-0

ICE Publishing: All rights reserved
http://dx.doi.org/10.1680/mc.61880.071

Appendix 2
Example quotations for compensation events

Section A: Based on the SCC

(Only used for main Options C, D and E.)

A2.1. Introduction

In this section we set out an example of a quotation based on the SCC for a hypothetical compensation event on the project named Spring Field.

The compensation event is for the provision of a new footbridge over the existing Spring Dyke following the realignment of Spring Road. For the purposes of this example, it is assumed that the new footbridge will be a new *section* of *works* (created by a supplemental agreement) to the existing contract for Spring Field.

We also assume for the purposes of this example that the contract has been let on an ECC Option C target contract with Activity Schedule.

This example sets out a format for the presentation of the quotation, and includes build-ups, supporting notes and comments on some of the issues surrounding the preparation of quotations for ECC4.

Figure A2.1 shows a sectional view of the proposed new footbridge. A programme for the *works* has been prepared, and is shown in Figure A2.2.

Figure A2.1 Proposed new footbridge over Spring Dyke

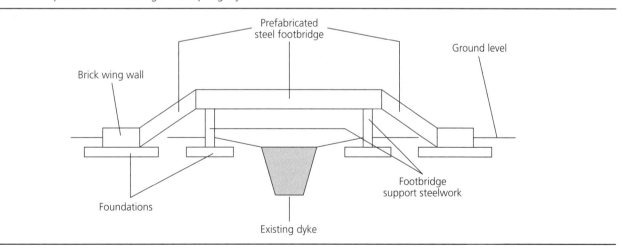

Managing Change

Figure A2.2 Programme for the new footbridge over Spring Dyke

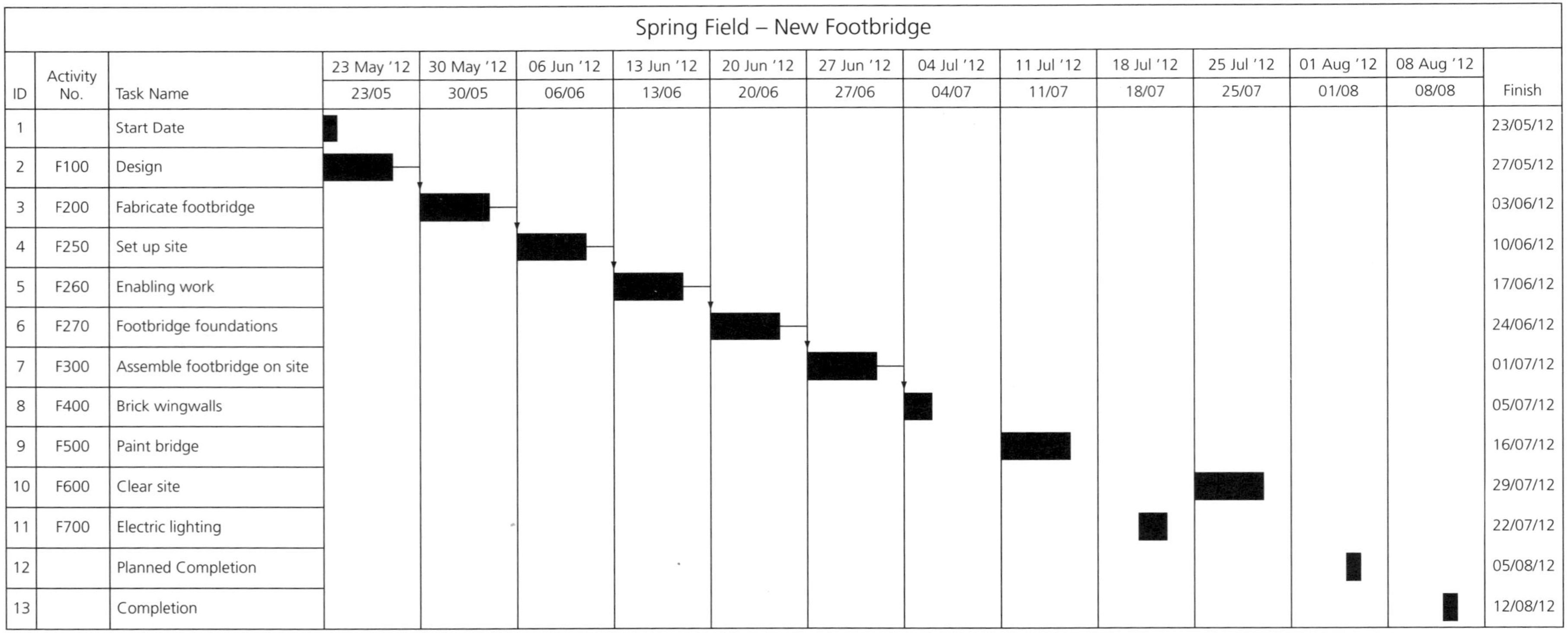

Spring Field – New Footbridge

ID	Activity No.	Task Name	23 May '12 23/05	30 May '12 30/05	06 Jun '12 06/06	13 Jun '12 13/06	20 Jun '12 20/06	27 Jun '12 27/06	04 Jul '12 04/07	11 Jul '12 11/07	18 Jul '12 18/07	25 Jul '12 25/07	01 Aug '12 01/08	08 Aug '12 08/08	Finish
1		Start Date	■												23/05/12
2	F100	Design	■												27/05/12
3	F200	Fabricate footbridge		■											03/06/12
4	F250	Set up site			■										10/06/12
5	F260	Enabling work				■									17/06/12
6	F270	Footbridge foundations					■								24/06/12
7	F300	Assemble footbridge on site						■							01/07/12
8	F400	Brick wingwalls							■						05/07/12
9	F500	Paint bridge								■					16/07/12
10	F600	Clear site										■			29/07/12
11	F700	Electric lighting									■				22/07/12
12		Planned Completion											■		05/08/12
13		Completion												■	12/08/12

Example quotations for compensation events

A2.2. Contract Data example

The following is an example of a quotation for the new footbridge.

Contract Data part two, extracted from the documents that are part of the contract used to produce this example quotation, is shown below.

A2.2.1 Contract Data part two – data provided by the *Contractor*

A2.2.1.1 Statements given in all contracts

- The *Contractor* is

Name	Virtual Contracting Limited
Address	Virtual House
	Virtual Lane
	Virtual City

- The *fee percentage* is 10%.
- The *working areas* are the Site and the area indicated on drawing FYK001 as lay-down and prefabrication facilities.
- The key people are

(1) Name	Joe Bloggs
Job	Site Agent
Responsibilities	....
Qualifications	....
Experience	....
(2) Name	John Public
Job	QS
Responsibilities	....
Qualifications	....
Experience	....

- The following matters will be included in the Early Warning Register:
  – flooding
  – unchartered services
- The Scope for the *Contractor's* design is in the document entitled '*Contractor's* Proposal'.
- The programme identified in the Contract Data is in the document entitled '*Contractor's* Programme'.

A2.2.1.2 Option C

- The *activity schedule* is in the document entitled 'Activity Schedule'.
- The tendered total of the Prices is £1843.00.

A2.2.1.3 Data for the SCC

- The listed items of Equipment purchased for work in the contract, with an on cost charge, are

Equipment	time-related on cost charge	per time period
Adjustable height restriction framework for vehicles	£100.00	per week

- The rates for special Equipment are

Equipment	rate
Bridge jack	£50 per week

- The rates for Defined Cost of manufacture or fabrication outside the Working Areas are

category of person	rate
Foreman	£20
Fabricators	£20

- The rates for Defined Cost of design outside the Working Areas are

category of person	rate
Draughtsman	£30

- The categories of design people whose travelling expenses to and from the Working Areas are included as a cost of design of the *works* and Equipment done outside of the Working Areas are

  None

A2.3. Defined Cost Defined Cost (clause 11.2(24)) is the cost of the components in the SCC less Disallowed Cost.

A2.4. Direct fee percentages The *fee percentage* covers such items as

- Head office charges and overheads:
  - loss of money insurance – loss of money due to theft
  - fidelity guarantee insurance – act of fraud or dishonesty
  - fire insurance – permanent premise and contents.
- Components not covered in the SCC include, for example:
  - insurance premiums,
    - professional indemnity insurance (*contractor*-designed work)
    - the *Client's* liability insurance
    - vehicle insurance
    - public liability insurance
    - all-risks insurance – loss or damage to permanent and/or temporary *works* and also covers Equipment and Plant and Materials (constructional plant)
    - corporation tax
  - advertising and recruitment costs
  - sureties and guarantees
  - indirect payments to staff on overseas contracts.
- The *Contractor's* profit.

Figure A2.3 shows the composition of Defined Cost.

Figure A2.3 Composition of Defined Cost

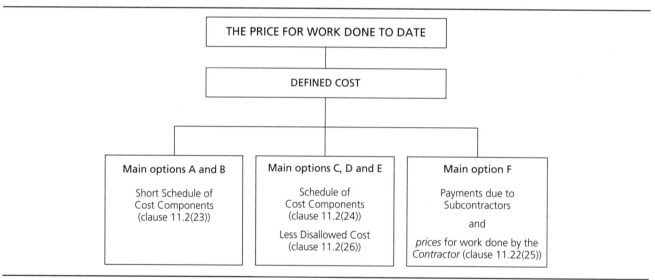

A2.5. SCC example quotation

We will now look at an example quotation for a compensation event for the ECC followed by an example of a Subcontractor's quotation.

A2.5.1 Defined cost

Table A2.1 Example quotation for a compensation event

CONTRACT: SPRING FIELD					
QUOTATION FOR COMPENSATION EVENT					
To: Project Manager				No: 41	
From: Virtual Contracting Limited				Sheet: 1	
Brief Description of Works: Provision of Footbridge to Spring Dyke Due to Road Realignment				Date: 1 July 20XX	
ACTIVITY NO./NOS: **F100 to F700**	DELAY TO PLANNED COMPLETION: **1 Day**			SECTION OF WORKS AFFECTED: **3A**	AFFECT ON KEY DATE: **Area 1–5 Days**
1 PEOPLE					
Activity F250 – Set up site (5 days)					
	No.	hours	Total hours	Rate	
Foreman	1	24	24	15.00	360.00
Labourers	4	40	160	10.51	1681.60
Ganger	1	40	40	8.00	320.00	2361.60
Activity F260 – Enabling Work (5 days)					
	No.	hours	Total hours	Rate	
Foreman	1	16	16	15.00	240.00
Labourers	3	40	120	10.51	1261.20	1501.20
Activity F270 – Footbridge Foundations (5 days)					
	No.	hours	Total hours	Rate	
Foreman	1	40	40	15.00	600.00
Labourers	3	50	150	10.51	1576.50
Ganger	1	40	40	8.00	320.00	2496.50
Activity F300 – Assemble Footbridge on site (5 days)					
	No.	hours	Total hours	Rate	
Foreman	1	50	50	15.00	750.00
Ganger	1	50	50	8.00	400.00
Steel fixers	4	50	200	12.00	2400.00
Crane operator	1	50	50	12.00	600.00
Banksman	1	50	50	11.00	550.00
Labourers	2	50	100	10.51	1051.00	5751.00
PEOPLE c/f					£12 110.30

CONTRACT: SPRING FIELD								
QUOTATION FOR COMPENSATION EVENT								
To: *Project Manager*							No: 41	
From: Virtual Contracting Limited							Sheet: 2	
Brief Description of Works: Provision of Footbridge to Spring Dyke Due to Road Realignment							Date: 1 July 20XX	
ACTIVITY NO./NOS: **F100 to F700**	DELAY TO PLANNED COMPLETION: **1 Day**						SECTION OF WORKS AFFECTED: **3A**	AFFECT ON KEY DATE: **Area 1–5 Days**
						PEOPLE b/f		£12 110.30
Activity F400 – Brick Wingwalls (2 days)								
	No.	Days	Total days	× hours	Total hours	Rate		
Foreman	1	2	2	10	20	15.00	300.00	
Bricklayers	4	2	8	10	80	12.00	960.00	
Labourer	2	2	4	10	40	10.51	420.40	1680.40
Activity P500 – Paint Bridge (5 days)								
	No.	Days	Total days	× hours	Total hours	Rate		
Painters	2	5	10	10	100	12.00	1200.00	1200.00
Activity P600 – Clear Site (5 days)								
	No.	Days	Total days	× hours	Total hours	Rate		
Foreman	1	5	5	8	40	15.00	600.00	
Labourer	2	5	10	10	100	10.51	1051.00	1651.00
PEOPLE c/f								£16 641.70

CONTRACT: SPRING FIELD			
QUOTATION FOR COMPENSATION EVENT			
To: *Project Manager*		No: 41	
From: Virtual Contracting Limited		Sheet: 3	
Brief Description of Works: Provision of Footbridge to Spring Dyke Due to Road Realignment		Date: 1 July 20XX	
ACTIVITY NO./NOS: **F100 to F700**	DELAY TO PLANNED COMPLETION: **1 Day**	SECTION OF WORKS AFFECTED: **3A**	AFFECT ON KEY DATE: **Area 1–5 Days**
		PEOPLE b/f	£16 64170
Activity F260 – Enabling Work (5 days) ■ People who are directly employed by the *Contractor* and whose normal place of work is not the Working Area but are working in the Working Areas Supervision of the erection of temporary works by Head Office – temporary works supervisor Team Leader 1 day		80.00	
■ People who are not directly employed by the *Contractor* but are paid for by him according to the time worked while they are in the Working Area – component 14 Inspection of the erected temporary works by Independent Third Party 1 day		150.00	230.00
PEOPLE TOTAL CARRIED TO SUMMARY =			£16 871.70

CONTRACT: SPRING FIELD			
QUOTATION FOR COMPENSATION EVENT			
To: Project Manager		No: 41	
From: Virtual Contracting Limited		Sheet: 4	
Brief Description of Works: Provision of Footbridge to Spring Dyke Due to Road Realignment		Date: 1 July 20XX	
ACTIVITY NO./NOS: **F100 to F700**	DELAY TO PLANNED COMPLETION: **1 Day**	SECTION OF WORKS AFFECTED: **3A**	AFFECT ON KEY DATE: **Area 1–5 Days**
2 EQUIPMENT			
Activity F250 – Set up site 2 Equipment Contractor's offices			
Section office – 1 No. at 10 m	6 weeks @ £30/week	180.00	
Contractor's site huts			
Store	6 weeks @ £20/week	120.00	
Site toilet	6 weeks @ £25/week	150.00	450.00
Activity F250 – Set up site (5 days) 21 Hired/Rented Equipment (at hire rate or rental rate multiplied by the time for which the Equipment is required)			
Dumper	5 days @ £50/week	50.00	
Lorry	5 days @ £100/day	500.00	
Shovel	say	10.00	
Hand-saw	say	20.00	580.00
Activity F260 – Enabling work (5 days)			
Dumper Hired Equipment	5 days @ £50/week	50.00	
Lorry	5 days @ £100/day	500.00	
Excavator	5 days @ £200/day	1000.00	1550.00
Activity F270 – Footbridge Foundations (5 days)			
Dumper Hired Equipment	5 days @ £50/week	50.00	
Concrete mixer	5 days @ £20/day	100.00	
Wheelbarrow	5 days @ £5/day	25.00	175.00
EQUIPMENT c/f			£2755.00

CONTRACT: SPRING FIELD			
QUOTATION FOR COMPENSATION EVENT			
To: *Project Manager*		No: 41	
From: Virtual Contracting Limited		Sheet: 5	
Brief Description of Works: Provision of Footbridge to Spring Dyke Due to Road Realignment		Date: 1 July 20XX	
ACTIVITY NO./NOS: **F100 to F700**	DELAY TO PLANNED COMPLETION: **1 Day**	SECTION OF WORKS AFFECTED: **3A**	AFFECT ON KEY DATE: **Area 1–5 Days**
		EQUIPMENT b/f	£2755.00
Activity F300 – Assemble Footbridge on Site (2 days) Dumper 2 days @ £10/day Lorry 1 day @ £100/day Setting out equipment say		20.00 100.00 40.00	160.00
22 Payments for Equipment which is not listed in the Contract Data but is owned, purchased or hired by the *Contractor* Hired Equipment Crane (incl. driver, delivery and removal from site) 2 days @ £50/day		50.00 100.00	100.00
Activity F400 – Brick Wingwalls (2 days) Dumper 2 days @ £10/day Hired Equipment Lorry 2 days @ £100/day Concrete mixer 1 day @ £40/day		20.00 200.00 40.00	260.00
Activity F500 – Paint Bridge (5 days) Hired Equipment Trestles 5 days @ £10/day		50.00	50.00
Activity F600 – Clear site (5 days) Dumper 7 days @ £10/day Hired Equipment Lorry 5 days @ £100/day		70.00 500.00	570.00
EQUIPMENT c/f			£3895.00

CONTRACT: SPRING FIELD			
QUOTATION FOR COMPENSATION EVENT			
To: Project Manager		No: 41	
From: Virtual Contracting Limited		Sheet: 6	
Brief Description of Works: Provision of Footbridge to Spring Dyke Due to Road Realignment		Date: 1 July 20XX	
ACTIVITY NO./NOS: **F100 to F700**	DELAY TO PLANNED COMPLETION: **1 Day**	SECTION OF WORKS AFFECTED: **3A**	AFFECT ON KEY DATE: **Area 1–5 Days**
	EQUIPMENT b/f		£3895.00
Activity F250 – Set up site 23 Payments for Equipment purchased for work in this contract listed with a time-related charge ■ Adjustable height restriction @ £100/week 4 weeks framework for vehicles		£400.00	£400.00
Activity F300 – Assemble footbridge 24 Payments for special Equipment listed in the Contract Data ■ Bridge Jacks 4 No. @ £50/week 4 weeks Additional item of special Equipment ■ Bridge deck lifting eyes @ £10/week 4 weeks		£800.00 £40.00	£840.00
Activity F250 – Set up site 25 Payments for the purchase of Equipment which is consumed ■ Sacrificial framework 10 sheets @ 20 ■ Fuel for Generator 10 litres @ 80p/litre		200.00 8.00	208.00
Activity F300 – Assemble footbridge 26 Unless included in the hire or rental rates, payments for ■ Transporting crane to and from site ■ Erecting and dismantling crane ■ Modifying jib of crane to lift bridge sections		100.00 100.00 500.00	700.00
27 Payments for purchase of materials used to construct or fabricate Equipment Materials for modifying jib of crane to lift bridge sections ■ Steel sections/RSAs ■ Lifting eye and chains		250.00 200.00	450.00
28 Unless included in the hire rates, the cost of operatives is included in the cost of people			
EQUIPMENT TOTAL CARRIED TO SUMMARY =			£6493.00

Example quotations for compensation events

CONTRACT: SPRING FIELD	
QUOTATION FOR COMPENSATION EVENT	
To: *Project Manager*	No: 41
From: Virtual Contracting Limited	Sheet: 7
Brief Description of Works: Provision of Footbridge to Spring Dyke Due to Road Realignment	Date: 1 July 20XX

ACTIVITY NO./NOS: **F100 to F700**	DELAY TO PLANNED COMPLETION: **1 Day**	SECTION OF WORKS AFFECTED: **3A**	AFFECT ON KEY DATE: **Area 1–5 Days**	
3 PLANT AND MATERIALS				
Activity F200 – Fabricate footbridge				
31 Purchasing Plant and Materials				
Materials for Footbridge				
■ Durasteel	100 m^2	@	£50	5000.00
■ RSA 45×45 mm	100 m	@	£5	500.00
■ Black bolts	500 No.	@	£2	1000.00
■ Stainless steel handrails	60	@	£100	6000.00	12 500.00
Activity F270 – Footbridge Foundations				
Concrete	80 m^3	@	£50	4000.00
Formwork	20 sheets	@	£20	400.00	4400.00
Activity F400 – Brick wing walls				
Bricks	2000 No.	@	£0.05p	100.00
■ Providing and removing packaging				
Providing storage boxes for the delivery of bridge bolts			£10.00	10.00
(Returned pallet)			(−£10)	(−10.00)
Disposal of polystyrene and cellophane wrapping around special bricks			50.00	50.00
■ Samples and tests				
Brick sample			£50	50.00
Paint sample board			£50	50.00
32 Cost is credited with payments received for disposal of Plant and Materials				
Surplus brick – restocked by stockist			Less £200.00	(−200.00)	50.00
PLANT AND MATERIALS TOTAL CARRIED TO SUMMARY =					£16 950.00

CONTRACT: SPRING FIELD			
QUOTATION FOR COMPENSATION EVENT			
To: *Project Manager*		No: 41	
From: Virtual Contracting Limited		Sheet: 8	
Brief Description of Works: Provision of Footbridge to Spring Dyke Due to Road Realignment		Date: 1 July 20XX	
ACTIVITY NO./NOS: **F100 to F700**	DELAY TO PLANNED COMPLETION: **1 Day**	SECTION OF WORKS AFFECTED: **3A**	AFFECT ON KEY DATE: **Area 1–5 Days**
4 SUBCONTRACTORS The following components of the cost of Subcontractors 41 Payments to Subcontractors for work which is subcontracted without taking into account any amounts paid to or retained from the Subcontractor by the Contractor, which would result in the *Client* paying or retaining the amount twice *Activity F700* – Electric lighting to footbridge Spring Electrics			7692.00
CHARGES TOTAL CARRIED TO SUMMARY =			£7692.00

CONTRACT: SPRING FIELD			
QUOTATION FOR COMPENSATION EVENT			
To: *Project Manager*		No: 41	
From: Virtual Contracting Limited		Sheet: 9	
Brief Description of Works: Provision of Footbridge to Spring Dyke Due to Road Realignment		Date: 1 July 20XX	
ACTIVITY NO./NOS: **F100 to F700**	DELAY TO PLANNED COMPLETION: **1 Day**	SECTION OF WORKS AFFECTED: **3A**	AFFECT ON KEY DATE: **Area 1–5 Days**
5 CHARGES *Activity F250* 51 Payments for provision and use in Working Areas of 1 Payment for temporary connection charges by Water Authority 2 Payment for temporary connection and disconnection of electrical supply to LEB		50.00 50.00	100.00
52 Payments to public authorities 1 Local Authority inspection charge		50.00	50.00
53 Payments for (a) to (h) e.g. 53(c) 1 Payment to Mr Jones for access to the Working Area and loss of crops		500.00	500.00
54 Payments made and received by the *Contractor* for the removal from Site and disposal and sale of materials from excavations and demolition		0.00	0.00
CHARGES TOTAL CARRIED TO SUMMARY =			£650.00

CONTRACT: SPRING FIELD	
QUOTATION FOR COMPENSATION EVENT	
To: *Project Manager*	No: 41
From: Virtual Contracting Limited	Sheet: 10
Brief Description of Works: Provision of Footbridge to Spring Dyke Due to Road Realignment	Date: 1 July 20XX

ACTIVITY NO./NOS: **F100 to F700**	DELAY TO PLANNED COMPLETION: **1 Day**	SECTION OF WORKS AFFECTED: **3A**	AFFECT ON KEY DATE: **Area 1–5 Days**	
6 MANUFACTURE AND FABRICATION				
6 The following components of cost of manufacture and fabrication of Plant and Materials, which are ■ Wholly or partly designed specifically for the works and ■ Manufactured or fabricated outside the Working Areas				
Activity F200 – Fabrication of footbridge				
61 The total of the hours worked by employees multiplied by the hourly rates stated in the Contract Data for the categories of employees listed				
Employee Costs				
Foreman	5 days × 8 hours	40	@	£20	800.00	
Fabricators, 8 men	5 days × 8 hours	320	@	£20	6400.00	7200.00
Subtotal						7200.00
MANUFACTURE AND FABRICATION TOTAL CARRIED TO SUMMARY =						**£7200.00**

CONTRACT: SPRING FIELD			
QUOTATION FOR COMPENSATION EVENT			
To: *Project Manager*		No: 41	
From: Virtual Contracting Limited		Sheet: 11	
Brief Description of Works: Provision of Footbridge to Spring Dyke Due to Road Realignment		Date: 1 July 20XX	
ACTIVITY NO./NOS: **F100 to F700**	DELAY TO PLANNED COMPLETION: **1 Day**	SECTION OF WORKS AFFECTED: **3A**	AFFECT ON KEY DATE: **Area 1–5 Days**
7 DESIGN *Activity F100 Design* 71 The total of the hours worked by employees multiplied by the hourly rates stated in the Contract Data for the categories of employees listed. (a) *Employee* Costs Draughtsman, 4 men × 10 hours per day × 5 days = 200 hours @ £30 72 The cost of travel to and from the Working Areas for the categories of design employees in the Contract Data (b) Travel costs (in Contract Data part two included in employee hourly rate)		6000.00 N/A	6000.00
DESIGN TOTAL CARRIED TO SUMMARY =			£6000.00
8 INSURANCE The following are deducted from cost: ■ the cost of events for which the contract requires the *Contractor* to insure and ■ other costs paid to the *Contractor* by the insurer		N/A N/A	
INSURANCE TOTAL CARRIED TO SUMMARY =			N/A

CONTRACT: SPRING FIELD			
QUOTATION FOR COMPENSATION EVENT			
To: *Project Manager*		No: 41	
From: Virtual Contracting Limited		Sheet: 12	
Brief Description of Works: Provision of Footbridge to Spring Dyke Due to Road Realignment		Date: 1 July 20XX	
ACTIVITY NO./NOS: **F100 to F700**	DELAY TO PLANNED COMPLETION: **1 Day**	SECTION OF WORKS AFFECTED: **3A**	AFFECT ON KEY DATE: **Area 1–5 Days**
SUMMARY			
1. PEOPLE			16 871.70
2. EQUIPMENT			6493.00
3. PLANT AND MATERIALS			16 950.00
4. SUBCONTRACTORS			7692.00
5. CHARGES			650.00
6. MANUFACTURE AND FABRICATION			7200.00
7. DESIGN			6000.0
8. INSURANCE			N/A
		Subtotal	**£61 856.70**
RISK ALLOWANCES			0.00
		Subtotal	**£61 856.70**
Fee 10%			6185.67
TOTAL DEFINED COST =			**£68 042.37**

A2.5.2 Defined Cost Subcontractor's work

Table A2.2 shows an example of the Spring Electrics quotation for their *works* in relation to the new footbridge. For the purposes of this example, we have assumed that the Subcontractor is on an Engineering and Construction Subcontract main Option C and has prepared a quotation using the SCC. It should be noted that it is highly likely the Subcontractor will submit their quotation in the form of a lump sum. It will be up to the *Contractor* to break the lump sum down to put into their own quotation.

Table A2.2 Example of a Subcontractor's quotation

CONTRACT: SPRING FIELD			
QUOTATION FOR COMPENSATION EVENT (Schedule of Cost Components)			
To: Virtual Contracting Limited		No: 25	
From: Spring Electrics		Sheet: 1	
Brief Description of Works: Electrics to new footbridge		Date: 25 June 20XX	
ACTIVITY NO./NOS: **G100**	DELAY TO PLANNED COMPLETION: **0 Days**	SECTION OF WORKS AFFECTED: **3A**	AFFECT ON KEY DATE: **Area 1–5 Days**
1 PEOPLE			
Activity F700 – Electric lighting to footbridge No. Days Total days hrs Total hrs Rate Foreman 1 2 2 10 20 20.00 Electrician 7 2 14 10 140 15.00 Trainee electrician 1 2 2 10 20 5.50		400.00 2100.00 110.00	
			£2610.00
2 EQUIPMENT			
Percentage adjustment for listed Equipment (Contract Data part two)……%			
EQUIPMENT TOTAL CARRIED TO SUMMARY =			£0.00
3 PLANT AND MATERIALS Lights 20 No. @ £100.00 Cabling 180m @ £10.00		2000.00 1800.00	3800.00
4 SUBCONTRACTORS			0.00
5 CHARGES			
CHARGES TOTAL CARRIED TO SUMMARY =			£0.00
6 MANUFACTURE AND FABRICATION			
MANUFACTURE AND FABRICATION TOTAL CARRIED TO SUMMARY =			£0.00
7 DESIGN			
DESIGN TOTAL CARRIED TO SUMMARY =			£0.00
8 INSURANCE Deduct from cost: ■ The cost of events which this contract requires the *Contractor* to insure and ■ Other costs paid by the insurer			
INSURANCE TOTAL CARRIED TO SUMMARY =			£0.00
9 RISK			
RISK TOTAL CARRIED TO SUMMARY =			£0.00

CONTRACT: SPRING FIELD		
QUOTATION FOR COMPENSATION EVENT (Schedule of Cost Components)		
QUOTATION FOR COMPENSATION EVENT (Schedule of Cost Components)		
To: Virtual Contracting Limited	No: 25	
From: Spring Electrics	Sheet: 1	
Brief Description of Works: Electrics to new footbridge	Date: 25 June 20XX	
ACTIVITY NO./ NOS: **G100**	DELAY TO PLANNED COMPLETION: **0 Days**	SECTION OF WORKS AFFECTED: **3A**	AFFECT ON KEY DATE: **Area 1–5 Days**

SUMMARY	
1. PEOPLE	2610.00
2. EQUIPMENT	0.00
3. PLANT AND MATERIALS	3800.00
4. SUBCONTRACTORS	0.00
5. CHARGES	0.00
6. MANUFACTURE AND FABRICATION	0.00
7. DESIGN	0.00
8. INSURANCE	0.00
9. RISK	0.00
TOTAL =	**£6410.00**
10. FEE *fee percentage* (see Contract Data part two) 20%	£1282.00
TOTAL DEFINED COST =	**£7692.00**

A2.6. Supporting notes

Supporting notes seeking to amplify and support the example quotation given in Section A2.2 above are now provided under the following headings:

- 'People – calculation of people (labour) costs'
- 'Example schedule of people rates from a quotations manual'
- 'Example calculation for percentage for Equipment purchased for work included in the contract'
- 'The *Contractor's* risk allowances'
- 'The Activity Schedule'
- 'Some reminders'.

A2.6.1 People – calculation of people (labour) costs

This is based upon (clause 63.1)

- the Defined Cost (as defined by the SCC) of the work already done
- the forecast Defined Cost of the work not yet done
- the resulting Fee.

The Fee (clause 11.2(8)) is defined as the sum of the amounts calculated by applying the *fee percentage* to Defined Cost. It is important to note that tender rates and prices are not used to assess change, and in all instances the cost for the different grades of people involved will be based on

- a payroll printout showing the required information
- proof of other payments such as lodging allowances
- other documentary evidence.

Table A2.3 Example of a *Contractor's* payroll

Payroll build-up based on the Schedule of Cost Components			Traditional calculation of hourly rates for labour based on the Working Rule Agreement	
General Operative Mr X for Period 1: 4-week period from 1 to 30 June 20XX			General Operative Mr X for Period 1: 4-week period from 1 to 30 June 20XX	
11	Wages and Salary *(Figure made up of Basic Rate, Additional Payments for skill, National Insurance and Training Levy Allowance from traditional build-up opposite)* [Items marked with an asterisk]	1161.21	Basic rate of pay (classification – General Operative, Skill Rate 1, 2, 3, 4, Craft Rate) 213 hours @ 4.78	1018.14*
		0.00	Additional payment for skilled work WRA (Schedule 1 – classification i, ii, iii) 0 hours @ 4.78	0.00*
12	Payments for			
(a)	Bonuses and incentives	213.00	Bonus – guaranteed minimum and production bonus 213 hours @ 1.00	213.00
(b)	Overtime	131.45	Non-productive overtime 27.5 hours @ 4.78	131.45
			Annual Wage (A)	*1362.59*
		incl. in 11	National Insurance employers contribution @ 10% of (A)	136.26
		incl. in 11	Training allowance or industrial training levy, e.g. CITB Training Levy 0.50% of PAYE (A)	6.81
(c)	Working in special circumstances	0.00	WRA Schedule 2 – Working in adverse conditions e.g. stone cleaning, tunnels, sewer work, working at height	0.00
(d)	Special allowances			
(e)	Absence due to sickness and holidays	94.92	Holiday credit 4 weeks @ 23.73	94.92
		27.25	Sick pay allowance @ 2% of (A) above	27.25
			Paid Total & Allowances (B)	*1627.83*
(f)	Severance related to work on this contract	24.42	Allowance for severance pay 1.5% of (B)	24.42
13	Payments in relation to people for			
(a)	Travel	0.00	WR.5 Travel allowances	0.00
(b)	Subsistence and lodging	508.48	WR.15 Subsistence (where applicable) 4 weeks × 7 nights × £18.16 per night	508.48
(c)	Relocation			0.00
(d)	Medical examinations			0.00
(e)	Passport and visas			0.00
(f)	Travel insurance			0.00
(g)	Items (a) to (f) for dependants			0.00
(h)	Protective clothing	8.31	Protective clothing 0.50% of (B)	8.13
(i)	Meeting the requirements of the law	32.56	Employers liability and public liability insurance 2% of (B)	32.56
(j)	Pensions and life insurance	0.00	Industry pension scheme	0.00
(k)	Death benefit	3.85	WR.21 Benefit schemes (Death benefit stamp) say	3.85
(l)	Occupational accident benefits			
(m)	Medical aid	33.37	Health Insurance say	33.37
(n)	A vehicle	0.00		
(o)	Safety training	0.00		
	Total	**2238.64**	**Total cost for period 1 = £**	**2238.64**
		incl. in fee percentage	Safety officer's time, QA Policy/inspection and all other costs and overheads, say 2.433% Note: These items are included in the calculation of labour costs using the Working Rule Agreement but are not a Component of People Costs in the ECC	54.47
	Total annual cost of general operative = £	2238.64	*Total annual cost of general operative =* £	2293.11
	Total hours worked = £	213.00	*Total hours worked =* £	213.00
	Cost per hour = £	10.51	*Cost per hour =* £	10.77

Table A2.3 shows a typical example of a *Contractor's* payroll printout and a build-up for the cost of people based on the components in the SCC.

In theory, this exercise based on the *Contractor's* accounts and records is needed for each and every category of people on each and every quotation for a compensation event. This principle is adequate where compensation events are few and far between, but on larger projects it may not be so practical.

Some *Clients* and *Contractors* have agreed on larger projects to do this exercise once a month or every quarter to establish a list of agreed rates to use for quotations. Some have developed what are called quotation manuals that are included at the time of tender to establish the first people costs and other costs for use to assess compensation events. This document is then reviewed every month, with a set of project-wide rates for the month or period ahead. This is particularly useful on large, multi-location projects that have large teams of people.

Although this is not strictly ECC policy, it is a practical way of providing consistency of approach across a project and reduces the need for too many people to have to get involved in establishing the Defined Cost of people, and reduces the fears of the *Contractor* in having to adopt an open-book approach on sensitive commercial information.

It also makes the audit of compensation events a great deal simpler.

A2.6.2 Example schedule of people rates from a quotations manual

Table A2.4 gives an extract from a quotations manual that shows the agreed hourly rates to be used in the assessment of compensation events during a month or period.

Table A2.4 Example schedule of people rates from a quotations manual

The following are the agreed rates for use on compensation events during the period 1–30 June 20XX

Staff	Grade/level	Hourly rate: £
Project Director	A	50.00
Agent	B	35.00
Site Foreman	C	30.00
QS	D	25.00
Engineer	E1	25.00
Engineer	E2	23.00
Engineer	E3	15.00
Planner	P	30.00
Buyer	B	20.00
Labour	Grade/level	Hourly rate: £
Ganger	S1	8.00
Bricklayer	S2	10.00
Labourer	S3	7.00[a]
Painter	S4	9.00
Foreman	S6	12.00
Steel fixer	S7	10.00
Crane operator	S8	8.00
Banksman	S9	7.00

[a] Rates used for forecast compensation events need to take into consideration future wages and material increases

Table A2.5 Schedule of Equipment purchased for the project – examples

Item	A – purchase price	B – period required	C – sale price	D – difference between purchase and sale price (A–C)	E – time-related on-cost charge (see Contract Data part two)
Dumper truck	£15 000.00	3 months	Say £13 000.00	£2 000.00	£20 per week for maintenance and repairs
Tunnel-boring machine	£150 000.00	12 months	Say £120 000.00	£30 000.00	£500 per month for maintenance and repairs

A2.6.3 Example calculation for Equipment purchased for work included in the contract

The ECC has a practical and realistic way of looking at Equipment purchased specifically for work on a project:

'Payments for Equipment purchased for work included in this contract listed with a time-related on cost charge, in the Contract Data, of

- the change in value over the period for which the Equipment is required and
- the time-related on cost charge stated in the Contract Data for the period for which the Equipment is required.' (SCC cost component 23)

A good example of this may be the purchase of a tunnel-boring machine, which could be new or refurbished, used for 12 months on a project. At the end of the project the tunnel-boring machine can be sold on to others.

An example calculation for cost component 23 for an item of Equipment purchased for the project is given in Table A2.5.

A2.6.4 *Contractor's* risk allowances

The contract states in clause 63.8 that

'The assessment of the effect of a compensation event includes risk allowances for cost and time for matters which have a **significant chance of occurring** and are not compensation events.'

It should be remembered that the *Contractor* carries all risks except those specifically taken by the *Client* in the Contract.

The key phrases here are matters that have '**significant chance of occurring**' and are '**not compensation events**'.

The conceptual idea is that the *Contractor* includes in quotations for compensation events for the 'risks' they carry under the contract in the same way as when they are tendering for the work.

The first exercise for the *Contractor* to undertake is to identify possible risks carried by them under the contract that have a significant chance of impacting the work that is the subject of the compensation event, including those transferred in new or amended contract conditions. Examples are potential or possible events on a particular project, such as boundary conditions, late flights, attempted suicides, eco-warriors and endangered species. The *Contractor* can make use of risk reduction meetings (clause 16.3) to discuss these matters with the *Project Manager*.

It therefore may be prudent for the *Contractor* or even the *Client* at an earlier stage to identify risks carried by the *Contractor*. It is suggested that such a list or pro forma could be suggested/included in the Scope so that it is clear from the outset how the *Contractor's* cost and time risk allowances are to be included and in what format they should be submitted with each quotation. An example of a *Contractor's* risk allowances schedule is given in Table A2.6.

In the example given in this appendix there may be no similar items in the original tender, therefore the rule should be to include for risks that have **a significant chance of occurring**.

Example quotations for compensation events

Table A2.6 Sample list of risk events that may need to be considered by the *Contractor* when preparing quotations

Contract title: Spring Field		Contract No: 2012/23		Quotation No: 41 – New Footbridge		
Item No.	Description	Probability (significance: low/medium/high)	How to include (in people rates, etc.)	Assessment details/assumptions	Impact assessment/time, cost	Mitigation details/time, cost
People						
1	Wage increases	New labour rates as of 1 July 20XX	Included in rates	Not applicable	None	Not applicable
2	Labour availability	Low	Not applicable	Not applicable	Not applicable	Not applicable
3	Subcontractor's availability	Low	Not applicable	Not applicable	Not applicable	Not applicable
4	Industrial relations	Low	Not applicable	Not applicable	Not applicable	Not applicable
5	Attendance on Subcontractors (welfare facilities)					
Equipment						
6	Equipment (constructional plant) increases	New rates as of 1 May 20XX	Included in rates	Not applicable	None	Not applicable
7	Equipment breakdown/maintenance	Low	Allowances made in output rates	Not applicable	Not applicable	Not applicable
8	Attendance on Subcontractors (Equipment, e.g. scaffold, cranes, etc.)	Low	Allowances made in output rates	Not applicable	Not applicable	Not applicable
Plant and Materials						
9	Plant and Materials availability (shortages, long lead-in times), delivery delays, etc.	High	Shortage of rebar for foundations	Built into the programme for the works	Not applicable	Buffer time built into the programme
Subcontractors						
10	Defective work	Low	Allowance made in the output rates	Build into the programme	Not applicable	Not applicable
11	Non-performance	Low	Allowance made in the output rates	Build into the programme	Not applicable	Not applicable
Manufacture and fabrication						
12	Manufacture and fabrication delays					
Design						
13	Design liability (increases in design liability) and professional indemnity (PI) insurance	Low	Design liability and PI already covered in main contract	Not applicable	Not applicable	Not applicable
14	Equipment design and temporary works	Low	Not applicable	Not applicable	Not applicable	Not applicable
15	Permanent design	Low	Not applicable	Not applicable	Not applicable	Not applicable
Workmanship/quality						
16	Workmanship/defective work/quality – setting out, etc.	Medium	Allowance made in output rates	Not applicable	Not applicable	Not applicable
17	Subcontractor's performance	Low	Not applicable	Not applicable	Not applicable	Not applicable
Method of working/constraints, etc.						
18	Method of working	Medium	Method of working may require adjacent road closure			

Managing Change

Contract title: Spring Field		Contract No: 2012/23		Quotation No: 41 – New Footbridge		
Item No.	Description	Probability (significance: low/medium/high)	How to include (in people rates, etc.)	Assessment details/ assumptions	Impact assessment/ time, cost	Mitigation details/ time, cost
19	Access restrictions	Low	Not applicable	Not applicable	Not applicable	Not applicable
20	Limitations of working space	Low	Not applicable	Not applicable	Not applicable	Not applicable
21	Existing overhead and underground services	Medium	Not applicable	Not applicable	Not applicable	Not applicable
22	Excesses in insurances	Low	Not applicable	Not applicable	Not applicable	Not applicable
23	Security of the Site (eco-warriors)	Low	Not applicable	Not applicable	Not applicable	Not applicable
24	Impact on future work (other sections of work, other packages, etc.)	Low	Not applicable	Not applicable	Not applicable	Not applicable
25	Output rates/productivity	Low	Not applicable	Not applicable	Not applicable	Not applicable
26	Health and safety					
Special safety requirements						
Client's risks in the contract						
27	Weather conditions	Low	If arises it will be a compensation event	Not applicable	Not applicable	Not applicable
28	Nature of the ground	Low	If arises it will be a compensation event	Not applicable	Not applicable	Not applicable
29	Working around other contractors	Low	If arises it will be a compensation event	Not applicable	Not applicable	Not applicable
Changes in the law (secondary Option X2)						
30	Secondary Option X2; if applicable, the *Client* takes the risk for changes in the law (landfill tax, employment law flexible hours, etc.)	Low	Secondary Option X2 is included in the contract. If it arises it will be a compensation event	Not applicable	Not applicable	Not applicable
Unforeseen risks						
31	Foot-and-mouth disease	Low	Not clear in the contract who has this risk			

It should be noted that where a *Project Manager* makes their own assessment, they should be making allowances for clause 63.8. In this way, it is visible how and what has been included in the quotation.

It would be prudent to call for this information when the *Contractor* submits their tender. Risk management should now become part of the assessment of a compensation event.

Therefore, if the *Contractor* can demonstrate that, for example, the ditch referred to has a significant chance of flooding, then the *Contractor* should be allowed to make allowance for this risk.

It is **not** the intention in the ECC for a blanket percentage to be added to each compensation event, as has been the temptation in some contracts; risk for each compensation event should be considered for each event. *Contractors* sometimes produce documents with all sorts of risks equating to a blanket percentage add-on to all compensation events, for example 25%. This is blatantly incorrect, and is a dangerous tactic for the *Contractor* since it means that they are not considering the risk issues properly on each compensation event, which should be of concern to both Parties and is a key principle of the ECC.

Table A2.7 Matters to be added to the Early Warning Register

Contract No: 2012/23			Quotation No: 41 – New Footbridge		
Matters					
Description	Probability (significance: low/medium/high)	How to include (in people rates, etc.)	Assessment details/ assumptions	Impact assessment/time, cost	Mitigation details/time, cost
Planning approval for the bridge delayed	Low	Not applicable	Approval received by 1 May 20XX	Delay to the project start date	Working closely with the *Project Manager* and local authority to gain planning permission. Looking at the possibility of pre-assembly off Site

Examples of *Contractor's* risk items are

- wage increases
- Plant and Material increases
- winter working (productivity outputs, etc.)
- Equipment hire rate increases
- a change in charges
- defective work
- maintenance time for Equipment (constructional plant).

In some instance the *Client* reallocates risks such as

- late flights
- road closures
- weather
- unforeseen ground conditions.

Care also needs to be taken to ensure that allowances are not duplicated: for example, allowance made in output rates in the programme and further allowances made in the rates and prices.

In the ECC the *Client* is required to list in Contract Data part one item 1, 'General', the matters to be included in the Early Warning Register (clause 11.2(8)). Likewise, the *Contractor* is required to do the same in Contract Data part two.

The intent of this Early Warning Register is to identify from the outset of the contract the potential risks associated with the contract. The risk is described, and also the actions to be taken to avoid or reduce the risk.

The Early Warning Register is about identifying and managing risk. Only compensation events allocate the risk between the parties.

Table A2.6 shows how a *Contractor* could show risk allowances associated with a compensation event. This example is by no means meant to be comprehensive. However, it does indicate how carefully both the *Contractor* and *Client* should consider risks associated with each compensation event. As well as identifying the specific risks for an event, it may also highlight new risks that should be identified on the contract Early Warning Register (Table A2.7).

A2.6.5 Activity Schedule Table A2.8 shows how the quotation for this compensation event may translate into new activities in the build-up to an Activity Schedule.

A2.6.6 Some reminders
- Cost and time effects of change are valued and adjusted collectively.
- Emphasis on pre-pricing/pre-assessment of compensation events using forecasts (Defined Cost as defined in the SCC) of work not yet done.

Table A2.8 Example Activity Schedule build-up

| Activity No | Description | Total: £ | Schedule of Cost Components ||||||||
			1 People	2 Equipment	3 Plant and Materials	4 Subcontractors	5 Charges	6 Manufacture and fabrication	7 Design	8 Insurance	Risk
F100	Design	6000.00							6000.00		
F200	Fabricate footbridge	19 700.00			12 500.00		650.00	7200.00			
F250	Set up the Site	4649.60	2361.60	1638.00							
F260	Enabling work	3281.20	1731.20	1550.00							
F270	Footbridge foundations	7071.50	2496.50	175.00	4400.00						
F300	Assemble footbridge on Site	8001.00	5751.00	2250.00							
F400	Brick wingwalls	1990.40	1680.40	260.00	50.00						
F500	Paint the bridge	1250.00	1200.00	50.00							
F600	Clear the Site	2221.00	1651.00	570.00							
F700	Footbridge lighting	7692.00				7692.00					
	Subtotals: £	61 856.70	16 871.70	6493.00	16 950.00	7692.00	650.00	7200.00	6000.00	0.00	0.00
	fee percentage 10%	6185.67									
	TOTAL DEFINED COST =	**£68 042.37**									

- Tendered rates and prices are not generally used to assess change (in all main Options rates and lump sums can be used by agreement).
- Costs of work already done based on Defined Cost.
- Assessments not revisited or adjusted when based on assumptions that are later corrected.
- Cost based on Defined Cost as defined in the SCC.
- Time based on entitlement, not need.
- Assessment to include *Contractor's* risk allowances (clause 63.8).
- A compensation event may create new risk matters that need to be included in the contract Early Warning Register.

Section B: Based on the SSCC

B2.1. Introduction The SSCC is used to assess compensation events for main Options A and B.

B2.1.1 Cost component 1 – people The people cost component has been simplified in item 11 to amounts calculated by multiplying each People Rate by the total time appropriate to that rate spent within the Working Areas.

Table B2.1 shows an extract of the people element of a quotation using the SSCC.

B2.1.2 Cost component 2 – Equipment This covers amounts for Equipment in the published list identified in Contract Data part two adjusted by the percentage adjustment listed in Contract Data part two. The published list, for example, will be the CECA Dayword Schedule, the RICS Dayword Schedule or such.

The entry in Contract Data part two is as follows:

- The published list of Equipment is the edition current at the Contract Date of the list published by CECA
- The percentage for adjustment for Equipment in the published list is −30%
- The rates for other Equipment are

  Equipment   rate

  ....

  ....

B2.1.3 Cost component 3 – Plant and Materials The same as the SCC.

Table B2.1 Example extract of the people element of a quotation using the SSCC

PEOPLE Activity F250 – Set up site (5 days)						
Category of people	No.	hrs	total hrs	rate		
Foreman	1	24	24	12.00	288.00	
Labourers	4	40	160	10.00	1600.00	
Ganger	1	40	40	7.00	280.00	2168.00
Activity F300, etc.						10 000.00
				Total for people	=	£12 168.00
				TOTAL FOR PEOPLE	=	£12 168.00

B2.1.4 Cost component 4 – Subcontractors   Same as the SCC.

B2.1.5 Cost component 5 – charges   Same as the SCC.

B2.1.6 Cost component 6 – manufacture and fabrication   This has been simplified to amounts paid by the *Contractor* (item 61 of the SSCC).

B2.1.7 Cost component 7 – design   The same as the SCC.

B2.1.8 Cost component 8 – insurance   The same as the SCC.

B2.1.9 Allowance *Contractor's* risk   The same as the SCC (see section A2.6.4 above).

Table B2.2 provides an example of a quotation using the SSCC for the *Contractor's* direct works only. Any subcontract element would be produced in an identical format by the Subcontractor using their tendered fee percentages. The *Contractor* adds their *subcontract fee percentage* to the amount of the Subcontractor's quotation.

Table B2.2 Example quotation using the SSCC

CONTRACT: SPRING FIELD				
QUOTATION FOR COMPENSATION EVENT **(Short Schedule of Cost Components)**				
To: *Project Manager*			No: 41	
From: Virtual Contracting Limited			Sheet: 3	
Brief Description of Works: Provision of Footbridge to Spring Dyke Due to Road Realignment			Date: 1 July 20XX	
ACTIVITY NO./NOS: **A500**	DELAY TO PLANNED COMPLETION: **0 Days**		SECTION OF WORKS AFFECTED: **3A**	AFFECT ON KEY DATE: **Area 1–0 Days**
1. PEOPLE				896.00
2. EQUIPMENT				560.00
3. PLANT AND MATERIALS				600.00
4. SUBCONTRACTORS				0.00
5. CHARGES				0.00
6. MANUFACTURE AND FABRICATION				0.00
7. DESIGN				0.00
8. INSURANCE				N/A
Subtotal				£2056.00
9. RISK ALLOWANCES				0.00
Subtotal (Defined Cost of other work – clause 11.2(8))				£2056.00
10. FEE			*fee percentage* 10%	205.60
TOTAL DEFINED COST =				**£2261.60**

Section C: Based on rates or lump sums – all main Options

For all main Options, if the *Project Manager* and *Contractor* agree, rates or lump sums can be used to assess compensation events (clause 63.2).

C3.1. Introduction

An example is provided in this section of a quotation where the *Project Manager* and *Contractor* have agreed that a compensation event for a proposed change to the Scope should be assessed using rates or lump sums (Table C2.1).

The inclusion of the use of rates and prices to assess compensation has been included to cover situations where it may not be practicable to use the SCC or SSCC to assess the change in the Prices. The compensation event is for the relatively small change to the Scope, and involves one additional trapped gully and 3 m of additional pipework to the drainage system (Figure C2.1). This change has been identified several weeks before the *works* are due to commence, and therefore there are no programme implications occurring due to this compensation event. For the purposes of this example, we assume that the contract has been let on ECC Option B priced contract with a *bill of quantities*.

Table C2.1 Example quotation using rates and lump sums

CONTRACT: WOOLLEY NEW ESTATE ROAD					
RATES AND LUMP SUMS					
CONTRACTOR'S QUOTATION					
To: Project Manager			CE No: 4.00		
From: Crackon Contractors			Sheet: 1 of x		
Brief Description of Works: Additional trapped gully required adjacent to new footpath to office block B as shown on drawing reference C123 R1 dated 1 July 20XX			Date: 1 July 20XX		
ACTIVITY NO./NOS: **500 Drainage works**	DELAY TO PLANNED COMPLETION: **None**		SECTION OF WORKS AFFECTED: **None**	KEY DATES AFFECTED: **None**	
In accordance with clause B63.13 (or A63.14, D63.13 dependent on the main Option selected) of the conditions of contract the Project Manager and Contractor have agreed to assess this compensation event using rates and lump sums. This is because the change is relatively small and the use of the Shorter SCC would be unduly lengthy in relation to the value of the compensation event.				Subtotal	total
BQ Reference	Description	Unit	Quantity	Rate	
I112.1	Vitrified clay pipes with flexible joints, etc., depth not exceeding 1.5 m	m	3	17.03	51.09
K320	Gullies, trapped, vitrified clay	m	1	241.83	241.83
				Total = £	322.21

Figure C2.1 Example drawing reference C123 R1 dated 1 July 20XX

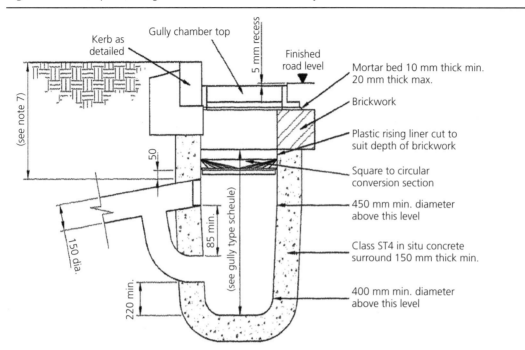

Plastic gully with in situ cast concrete surround

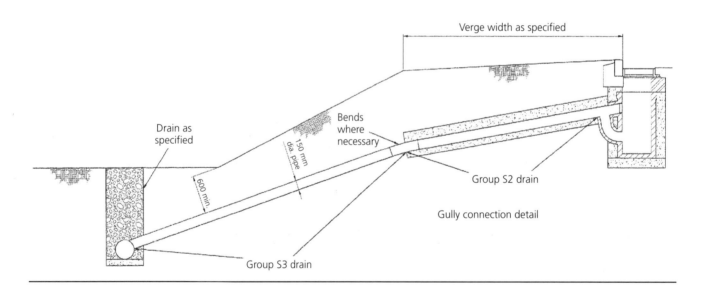

Gully connection detail

Managing Change
ISBN 978-0-7277-6188-0

ICE Publishing: All rights reserved
http://dx.doi.org/10.1680/mc.61880.099

Appendix 3
Example people cost calculations

A3.1. Introduction   This appendix looks at the comparison of how people (labour costs) would be calculated traditionally by contractors, and how this compares with the SCC.

It also shows how these costs interact with each other and how they can be calculated and dealt with.

Table A3.1 compares side by side the SCC and a traditional build-up. It highlights the differences between the two calculations and indicates how these differences in the costs are to be dealt with when using the SCC.

Table A3.1 Calculation of people costs – side-by-side comparison of the SCC and a traditional calculation of labour costs

Rate build-up based on Schedule of Cost Components			Traditional calculation for labour costs Annual cost of wages (Based on Construction Industry Joint Council – Working Rule Agreement)			
11	**Wages and salary** (Figure made up of Basic Rate, Additional Payments for skill, National Insurance and Training Levy Allowance from traditional build-up opposite)	14 723.10	Basic rate of pay (Classification – General Operative, skill rate 1, 2, 3, 4, Craft Rate) Additional payment for skilled work WRA Schedule 1 – Classification i, ii, iii	2694 hrs @ 4.78 0 hrs @ 4.78	12 877.32 0.00	
12	**Payments for**					
(a)	bonuses and incentives	2694.00	Bonus – guaranteed minimum and production bonus	2694 hrs @ 1.00	2694.00	
(b)	overtime	2007.60	Non-productive overtime	420 hrs @ 4.78	2007.60	
			Annual wage (A)		17 578.92	
		incl. in 11	National Insurance, employer's contribution @ 10% of (A)		1757.89	
		incl. in 11	Training Allowance or Industrial Training Levy, e.g. CITB Training Levy 0.50% of PAYE (A)		87.89	
(c)	working in special circumstances	0.00	WRA Schedule 2 – Working in adverse conditions, e.g. stone cleaning, tunnels, sewer work, working at height		0.00	
(d)	special allowances					
(e)	absence due to sickness and holidays	1230.85 351.58	Holiday credit Sick pay allowance @ 2% of (A) above	52 wks @ 23.67	1230.85 351.58	
			Paid total and allowances (B)		21 007.13	
(f)	severance related to work on this contract	315.11	Allowance for severance pay 1.5% of (B)		315.11	
13	**Payments in relation to people for**					
(a)	travelling	0.00	WR.5 Travel allowances		0.00	
(b)	subsistence and lodging	5974.64	WR.15 Subsistence (where applicable) 47 weeks × 7 nights × £18.16 per night		5974.64	
(c)	relocation				0.00	
(d)	medical examinations				0.00	
(e)	passport and visas				0.00	
(f)	travel insurance				0.00	
(g)	items (a) to (f) for dependants				0.00	
(h)	protective clothing	105.04	Protective clothing 0.50% of (B)		105.04	
(j)	contributions, levies or taxes imposed by law	420.14	Employer's liability and public liability insurance 2% of (B)		420.14	
(k)	pension and life insurance	0.00	Industry pension scheme		0.00	
(l)	death benefit	50.00	WR.21 Benefit schemes (death benefit stamp)	say	50.00	
(m)	occupational accident benefits					
(n)	medical aid and health insurance	452.51	Health insurance	say	452.51	
(o)	a vehicle	0.00	Vehicle (assumed dealt with separately)			
(p)	safety training	0.00	Safety Training (demonstration on ongoing commitment to safety)			
	Total annual cost =	£ 28 324.57	**Total annual cost =**	£	28 324.57	
	These items need to be included in the *fee percentage*		Safety officer's time, QA Policy/inspection and all other costs and overheads say 2.5% Note: These items are included in the calculation of labour costs using the Working Rule Agreement but are not a Component of People Costs in the full SCC		708.11	
	Total annual cost of general operative =	£ 28 324.57	Total annual cost of general operative =	£	29 032.68	
	Total hours worked =	£ 2694.00	Total hours worked =	£	2694.00	
	Cost per hour =	£ 10.51	Cost per hour =	£	10.78	

Managing Change
ISBN 978-0-7277-6188-0

ICE Publishing: All rights reserved
http://dx.doi.org/10.1680/mc.61880.101

Appendix 4
Comparison between traditional preliminaries build-up and how they relate to the Schedule of Cost Components and the Short Schedule of Cost Components

A4.1. Introduction

This appendix aims to clarify by way of comparison how a traditional preliminaries build-up and the ECC relate to each other. In particular, a traditional example build-up of preliminaries is given. This is then compared to the SCC and the SSCC to show how they are dealt with (Table A4.1).

The key changes to the SCC and SSCC in NEC4 compared with NEC3 are

- People:
  - people whose normal place of working is not the Working Area are paid proportionally to the time they spend in the Working Area.
- Subcontractors:
  - Subcontractors are now a cost component
  - payments to Subcontractors for work that is subcontracted without taking into account any amounts paid to or retained from the Subcontractor by the *Contractor*, which would result in the *Client* paying or retaining the amount twice.
- Charges:
  - Defined Cost is now actual Defined Cost and there is no longer a Working Area Overhead applied to the cost of people.

Some items that were deemed to be in the Working Area Overhead are not included as items of actual Defined Cost within the other cost components.

Table A4.1 Comparison between a traditional tender build-up for staff, Site on-costs and other items and how and where they are included in the SCC and the SSCC (numbers denote the cost component)

Item description			£	ECC	
				SCC	SSCC
Contractor's Site on-costs – time-related					
Site staff salaries					
Agent	30 weeks	£450	13 500.00	People	People
Senior engineer	30 weeks	£380	11 400.00	People	People
Engineers	30 weeks	£300	9000.00	People	People
General foreman	30 weeks	£400	12 000.00	People	People
Office manager	30 weeks	£400	12 000.00	People	People
Timekeeper	30 weeks	£240	7200.00	People	People
Storeman/checker	30 weeks	£200	6000.00	People	People
Typist/telephonist	30 weeks	£150	4500.00	People	People
Cost clerk	30 weeks	£240	7200.00	People	People
Quantity surveyor	30 weeks	£450	13 500.00	People	People
Fitter					
Site staff expenses (1% of staff salaries)				People	People
Attendant labour					
Chainman	30 weeks	£190	5700.00	People	People
Driver	30 weeks	£220	6600.00	People	People
Office cleaner (part-time)	30 weeks	£80	2400.00	People	People
General yard labour					
(loading and offloading, clearing site, rubbish, etc.)					
Ganger – 1 No.	30 weeks	£220	6600.00	People	People
Labourer – 2 No.	30 weeks	£220	6600.00	People	People
Plant Maintenance (Contractors's own constructional plant)					
Equipment in the ECC					
Fitter	30 weeks	£350	10 500.00	People (note Equipment 28)	People (note Equipment 28)
Fitters mate	30 weeks	£300	9000.00	People (note Equipment 28)	People (note Equipment 28)
Total carried forward =		£	143 700.00		

Appendix 4

Item description			£	ECC	
				SCC	SSCC
Total brought forward			143 700.00		
Contractor's Site on-costs – time-related					
Site transport – staff					
Agent's car	30 weeks	£100	3000.00	Equipment/People 13(n)	Equipment/People 11
Engineer's car	30 weeks	£90	2700.00	Equipment/People 13(n)	Equipment/People 11
4 × 4 utility car	30 weeks	£150	4500.00	Equipment	Equipment
(for general use on Site)					
Site transport – labour					
Bus – two trips per day	30 weeks	£250	7500.00	Equipment	Equipment
Contractor's offices					
Mobile office	30 weeks	£120	3600.00	Equipment	Equipment
(10 staff × 8 m^2 = 80 m^2)					
Section offices (1 No. at 10 m^2)	30 weeks	£30	900.00	Equipment	Equipment
Contractor's site huts					
Stores hot 30 m^2	30 weeks	£20	600.00	Equipment	Equipment
Canteen 70 m^2	30 weeks	£100	3000.00	Equipment	Equipment
Washroom 30 m^2	30 weeks	£25	750.00	Equipment	Equipment
Staff toilets	30 weeks	£30	900.00	Equipment	Equipment
Site toilets	30 weeks	£30	900.00	Equipment	Equipment
Rates				Nil	Equipment	Equipment
General					
General office expenditure	30 weeks	£50	1500.00	Charges	Charges
Telephone calls and rental	30 weeks	£50	1500.00	Charges 51	Charges 51
Furniture rental	30 weeks	£35	1050.00	Equipment	Equipment
Canteen and welfare rental	30 weeks	£40	1200.00	Equipment	Equipment
Surveying equipment rental	30 weeks	£30	900.00	Equipment	Equipment
Testing equipment rental	30 weeks	£25	750.00	Equipment	Equipment
Lighting and heating offices	30 weeks	£45	1350.00	Charges 51	Charges 51
Water supply	30 weeks	£20	600.00	Charges 51	Charges 51
Small tools (1% of labour cost)		say	5000.00	Equipment	Equipment
Protective clothing		say	2500.00	People	People
(0.5% of labour cost)					
Road lighting					
Cleaning roads					
Road sweeper	30 weeks	£50	1500.00	Equipment	Equipment
Labour	30 weeks	£150	4500.00	People	People
Progress photographs		say	1000.00	Equipment	Equipment
Total Contractor's Site on-costs – time-related =		£	195 400.00		

Item description		£	ECC	
			SCC	SSCC
Contractor's Site on-costs – non-time-related				
(These costs are unlikely to be affected by compensation events)				
Erect and dismantle offices				
Mobile	say	500.00	Equipment	Equipment
Site offices	say	500.00	Equipment	Equipment
Toilets	say	500.00	Equipment	Equipment
Wiring, water, etc.	say	500.00	Equipment	Equipment
Erect and dismantle other buildings				
Stores	say	500.00	Equipment	Equipment
Welfare	say	500.00	Equipment	Equipment
Toilets	say	500.00	Equipment	Equipment
Wiring, water, etc.	say	500.00	Equipment	Equipment
Telephone installation	say	500.00	Charges 51	Charges 51
Furniture and office equipment				
Purchase cost less residual value	say	500.00	Equipment	Equipment
Survey equipment and setting out				
Purchase cost, pegs, paint, profiles, etc.	say	700.00	Equipment 25	Equipment 25
Canteen and welfare equipment				
Purchase cost less residual value	say	1500.00	Equipment	Equipment
Electrical installation				
Water supply				
Connection charges	say	500.00	Charges 51	Charges 51
Site installation	say	500.00	Charges	Charges 51
Transport of plant and equipment	say	500.00	Equipment	Equipment
Stores compounds and huts	say	500.00	Equipment	Equipment
Sign boards and traffic signs	say	500.00	Equipment	Equipment
Insurances				
Contractors ALL risk (1.5% on £500 000)		7500.00	fee %	fee %
Allow for excesses		200.00	fee %	fee %
General Site clearance	say	1500.00	People/Equipment	People/Equipment
Total Contractor's Site on-costs – non-time-related = £		**20 700.00**		

Appendix 4

Item description				£	ECC	
					SCC	SSCC
Client's and Consultant's requirements on Site – time-related						
(Details of the requirements will be defined in the Scope)						
Attendant labour	30 weeks	£100.00		3000.00	People	People
Offices 40 m^2	30 weeks	£100.00		3000.00	Equipment	Equipment
Transport 2 4 × 4 utility cars	30 weeks	£300.00		9000.00	Equipment	Equipment
Telephone calls and rental	30 weeks	£25.00		750.00	Charges 51	Charges 51
Furniture and Equipment	30 weeks	£25.00		750.00	Equipment	Equipment
Survey equipment	30 weeks	£25.00		750.00	Equipment	Equipment
Heating and lighting 40 m^2	30 weeks	£30.00		900.00	Charges 51	Charges 51
Office consumables	30 weeks	£20.00		600.00	Charges 53(h)	Charges 53(h)
Office cleaning	30 weeks	£20.00		600.00	People/Subcontractor	People/Subcontractor
Total Client's and Contractor's requirements – time-related =			£	**19 350.00**		

Item description		£	ECC	
			SCC	SSCC
Client's and Consultant's requirements on Site – non-time-related				
(Details of the requirements will be defined in the Scope)				
Erection and dismantling of engineer's offices say		500.00	Equipment 26	Equipment 26
Toilets say		250.00	Equipment	Equipment
Telephone installation	say	200.00	Charges 51	Charges 51
Electrical installation	say	200.00	Charges 51	Charges 51
Furniture and Equipment				
Purchase price less residual value	say	2000.00	Equipment	Equipment
Progress photographs say		250.00	Equipment 25	Equipment 25
Total Client's and Consultant's requirements – time-related =	£	**3400.00**		

Item description			£	ECC	
				SCC	SSCC
Temporary works not included in unit rates					
Temporary fencing					
500 m chestnut fencing – materials	500 m	£10	5000.00	Plant and Materials	Plant and Materials
Plant		say	1000.00	Equipment	Equipment
Labour	200 man-hours	£10	2000.00	People	People
Traffic diversions		say	2000.00	People/Plant and Materials/Equipment	People/Plant and Materials/Equipment
Footpath diversion		say	500.00	People/Plant and Materials/Equipment	People/Plant and Materials/Equipment
Site access roads		say	4000.00	People/Plant and Materials/Equipment	People/Plant and Materials/Equipment
Total temporary works not included in unit rates =		£	**14 500.00**		

Item description			£	ECC	
				SCC	SSCC
General purpose constructional plant (Equipment) and plant not included in unit costs					
Wheeled tractor					
Hire	30 weeks	£240	7200.00	Equipment (21 externally hired or 22 internally hired)	Equipment (published schedule or list in Contract Data)
Driver	30 weeks	£1250	37 500.00	People (28 unless included in hire rates or depreciation and maintenance charge)	People (26 unless included in hire rates)
Consumables	30 weeks	£50	1500.00	Equipment 25	Equipment 25
Crane 20 m, 15 tonne					
Hire	30 weeks	£250	7500.00	Equipment	Equipment
Driver	30 weeks	£1250	37 500.00	People	People
Consumables	30 weeks	£50	1500.00	Equipment	Equipment
Sawbench					
Hire	30 weeks	£30	900.00	Equipment	Equipment
Consumables	30 weeks	£10	300.00	Equipment	Equipment
Compressor 21 m^3 silenced					
Hire	30 weeks	£200	6000.00	Equipment	Equipment
Consumables	30 weeks	£20	600.00	Equipment	Equipment
Pumps					
Hire	30 weeks	£50	1500.00	Equipment	Equipment
Consumables	30 weeks	£20	600.00	Equipment	Equipment
Total general purpose constructional plant not included in unit costs =		£	102 600.00		

Managing Change
ISBN 978-0-7277-6188-0

ICE Publishing: All rights reserved
http://dx.doi.org/10.1680/mc.61880.107

Appendix 5
Interrelationship between the *Contractor's* and the Subcontractor's *share* on target cost contracts

A5.1. Introduction   This appendix aims to clarify by way of an example the interrelationship between the *Contractor's* and the Subcontractor's *share* on target cost contracts.

Table A5.1 shows how the *Contractor's* share is calculated.

Table A5.1 Interrelationship between the *Contractor's* and the Subcontractor's share on target contracts (assuming that the Price for Work Done to Date is less than the total of the Prices)

Main Contractor ECC Option C or D Target Cost Contract		Subcontractor A ECC Subcontract Option C or D Target Cost Contract	
***Contractor's share* 50/50 of the difference between the total of the Prices (target) and the Price for Work Done to Date (Defined Cost)**		***Subcontractor's share* 50/50 of the difference between the total of the Prices (target) and the Price for Work Done to Date (Defined Cost)**	
		Subcontractor's share 50% of under- or overspend	
Tendered total of the Prices (target cost)	£220 000.00	Tendered total of the Prices (target cost)	£200 000.00
(£200 000.00 from Subcontractor plus *fee percentage* of 10% = £220 000.00)			
Price for Work Done to Date (Defined Cost) (including Fee)	£192 500.00	Price for Work Done to Date (Defined Cost) (including Fee)	£150 000.00
(£175 000.00 from Subcontractor plus *Contractors's fee percentage* of 10% = £192 500)			
Difference (saving)	(A) £27 500.00	Difference (saving)	(A) £50 000.00
Calculation for the Defined Cost		**Calculation for the Defined Cost**	
Price for Work Done to Date (Defined Cost)	£192 500.00	Price for Work Done to Date (Defined Cost)	£150 000.00
Contractor's share (Difference between total of the Prices and the Price for Work Done to Date) from (A) above		***Subcontractor's share*** (Difference between total of the Prices and the Price for Work Done to Date) from (A) above	
£27 500.00 × 50%	£13 750.00	£50 000 × 50%	£25 000.00
Total paid to *Contractor* = (by *Client* to *Contractor*)	£206 250.00	**Total paid to Subcontractor =** (by *Contractor* to Subcontractor)	£175 000.00

Managing Change
ISBN 978-0-7277-6188-0

ICE Publishing: All rights reserved
http://dx.doi.org/10.1680/mc.61880.109

Index

Abbreviations used:
CE, compensation event; SCC, Schedule of Cost Components; SSCC, Short Schedule of Cost Components.

acceleration, quotation for, 8, 23, 69
acceptance of quotations *see* quotations
Accepted Programme, 7, 13, 18, 23, 25, 26, 27
access (to Site), 69
  Client not allowing access to and use of each part of Site, 7
access date, 7
activity schedule and Activity Schedule, 18, 24, 45, 46, 48, 66, 67
  build-up, 94
  in example quotation for CE, 71, 73, 87, 93–95
additional CEs, 6, 14–15, 17, 22, 32
additional reimbursement, 9
Adjudicator and adjudication, 22, 26, 29, 33
advanced payment to *Contractor*, 17
ammending contracts prior to execution, 31
assessment (evaluation/valuation/administration)
  CEs, 5, 18–20, 26–28, 45–46, 46, 46–47, 68
  of changes, 45
    to prices, 18–19, 23, 45, 64, 65
  quotations, 26–27, 68
assessment date, 63
attendant labour, 102
audits, 63–64

Bill of Quantities (and *bill of quantities*), 15–16, 17, 24, 45, 46, 48, 67, 97
breach of contract, 13, 30
build up
  Activity Schedule, 94
  payroll, 88
  preliminaries, 70, 101–106
  Rate, 100
building information management, 17

charges as cost component
  in example quotation for CE, 82, 101
  SCC, 54, 82
    in final calculation for presentation for amount due for payment, 62
  SSCC, 59, 96
    in final calculation for presentation for amount due for payment, 62
claims, 29

Clients
  liability in *conditions of contract*, 11
  not allowing access to and use of each part of Site, 7
  not providing material/facilities/samples for tests and inspection, 13
  on-site requirements, 105
  responsibilities, 9
    work to be done, 7
  risks, 92
collaborative management, 4
communications, not replying to, 7–8, 38
compensation events (CEs), 3–42, 67, 68, 71–98
  additional, 6, 14–15, 17, 22, 32
  believing an event is a CE, 21, 37
  complicating factors, 18, 42
  correction to an assumption about, 3, 21
  evaluation/valuation/assessment/administration, 5, 18–20, 26–28, 45–46, 46–47, 48, 50, 56, 68
    programme in, 27–28
    rates and prices in, 97
  financial effects *see* financial effects
  frequently asked questions, 29–33
  grouping, 30–31
  implementation, 28
  list of events and where to find them, 6–14
    summary, 15
  meaning, 5–6
  notification, 20–22
  numerous small, 68–69
  procedure, 35–42
    background, 4–5
    flow charts, 18, 38–41
  quotation *see* quotation
  removing, 31–32
  revisiting, 19
  roles of *Project Manager* and *Contractor* relating to, 18
  SSC/SSCC and, 45–46, 46–47, 48, 50, 56, 63, 65, 67, 68–69
Completion, 13
  delays, 17, 25
  in example quotation for CE, 75–87
  unpreventable events not stopping completion of whole of *works*, 13–14
Completion Date (and changes to it), 18, 22
  changes, 18

Completion Date (and changes to it) (*continued*)
  delays, 23, 25, 33
conditions of contract, *Clients* liability in, 11
constructing as cost component, SCC, 53
Consultant's requirements on Site, 105
consumables
  SCC, 53, 62
  SSCC, 58, 59
contra proferentem rule, 9
contract(s), amending prior to execution, 31
contract(s)
  breach of, 13, 30
  cost-based contract *see* cost-based contracts
  cost-reimbursable-type, 31, 55
  lump sum contract *see* lump sum
  management contract, 48
  payment during, 45, 46, 46–47
  payment for Equipment purchased for work included in, 52, 60, 73, 90
  payroll (payments to people), 88, 100
  priced-based contract, 45–46, 48
  target cost contract, 5, 46, 107–108
Contract Data, 2
  CEs and, 5, 7, 8, 10, 11, 13, 14, 15, 22, 25
  SCC and/or SSCC, 44, 48, 49, 50, 51, 53, 54, 55, 56, 57, 58, 59, 61, 63, 64, 65, 68–69, 90, 93, 95
    Equipment not listed in, 52, 58, 60, 65, 79
    example, 73–74
    payments for Equipment, 52, 53, 60, 61, 65, 73, 79, 80
Contract Date, 9, 13, 15, 16, 17, 74, 95
Contractors
  actions required with CEs, 38–39
  assessment of CE by, 26
  limitation of liability (for their design), 17
  management *Contractor*, 48
  notification of CEs by, 21–22, 37, 38
  offices, 103
  payments to *see* payments
  people directly employed by, 51, 54, 60, 65, 77
  risk *see* risk
  roles (relating to CEs), 18
  share, target cost contracts, 107–108
cost, 43–70
  components *see* Schedule of Cost Components; Short Schedule of Cost Components
  Defined *see* Defined Cost
  direct, 54, 59
  Disallowed, 25, 49, 54, 59, 70, 74
  unit, plant not included in, 105
cost-based contracts, 46–47, 48
  cost-reimbursable-type contracts, 31, 55
  target cost contracts, 5, 46, 107–108

dates
  Contract Date, 9, 13, 15, 16, 17, 74, 95
  defects date, 8, 29
  Key Dates *see* Key Dates
  Price for Work Done to Date, 17, 25, 45, 46, 47, 48, 64, 108

  see also time issues and timescales
decisions, changed, 8, 20, 28, 36
Defect(s)
  CEs and, 8
  corrected, 64, 69
  uncorrected, 69
defect correction period, 69
defects date, 8, 29
Defined Cost, 16, 17, 18, 19, 24, 25, 28, 30, 48–49, 68
  definition, 48–49
  occasions when not used, 68
  SCC and/or SSCC, 44, 45, 48–49, 50, 52, 54, 55, 56, 58, 59, 63, 65, 68–69, 101
    in example quotation for CE, 71, 74, 75–85, 86, 87, 89, 93, 94, 95, 96
    subcontractor's work, 86
    target cost contracts, 108
delays, 17
  to Completion *see* Completion; Completion Date
  to Key Dates, 25
  tests and inspections causing unnecessary delay, 8
design
  as cost component
    in example quotation for CE, 84
    SCC, 55, 84
    SSCC, 59, 96
  limitation of *Contractor's* liability for, 17
  risk relating to, 91
direct cost, 54, 59
Disallowed Cost, 25, 49, 54, 59, 70, 74
dismantling as cost component
  SCC, 53, 104
  SSCC, 58, 104
drawings, Site Information, 108
duties *see* roles and responsibilities

early warnings, 9, 23, 24, 30
  Early Warning Register, 93, 95
eight weeks since becoming aware of a CE, 4, 21–22, 29
employees *see* people
Engineering and Construction Contract (basic references), 1–2
Equipment
  as cost component, 51–53, 60–61, 78–80
    in example quotation for CE, 78–79
    SCC, 51–53, 60–61, 65, 78–80, 102
    SSCC, 57–58, 61, 65, 102
  risk relating to, 91
erection as cost component
  SCC, 53, 104
  SSCC, 58, 104
evaluation *see* assessment
execution, amending contracts prior to, 31
expenses, site staff, 102

fabrication *see* manufacture and fabrication
facilities, Materials, *Client* not providing them for inspection and tests, 13

Fee, 18, 25, 49, 65
  definition, 25, 49
fee percentage, 18, 25, 49, 50, 51, 56, 66, 70, 74
  direct, 56, 67, 74
  subcontract, 96, 108
financial effects of CEs (CEs), 48
  evaluation/assessment, 45–46, 48, 50, 56
  forecast of CEs, 5, 19
five-point test, 21, 22, 23
force majeure, 14
forecasts
  of financial effects of CEs, 5, 19
  of work not yet done, 24–25

general yard labour, 102

hire of Equipment, 52
historical object of interest, 8, 20
hourly labour rates, 89
  traditional calculation, 88
Housing Grants, Construction and Regeneration Act (1996), 17
huts, 103

Information Execution Plan, 17
inspection, 69
  Client not providing material/facilities/samples for, 13
  delay due to, 8
instructions (for submitting quotations with CEs), 20–22, 23, 26, 28, 36
insurance as cost component
  SCC, 55, 65, 84
    in example quotation for CE, 84
  SSCC, 60, 65, 96

Key Dates, 25
  CEs and, 5, 7, 17, 18, 21, 22, 23, 25
    in example quotation, 75–87
  delay to, 25

labour
  attendant, 102
  costs *see* people
  general yard, 102
  hourly rates *see* hourly labour rates
law (legal matters)
  changes in, 17, 92
  Contractor notifying CE more than 8 weeks after becoming aware of it, 29
  force majeure, 13
  legal matters *see* law
liability
  Clients, in *conditions of contract*, 11
  Contractor's, limitation (for their design), 17
lump sums, 31, 45, 46, 48, 86, 97
  'min-lump sum', 19

main Options, 15–17, 44, 45, 47, 48, 49, 51, 56, 64, 66, 74, 95

CEs, 15–17
  in rates and lump sum example quotation, 97
management *Contractor*, 48
manufacture and fabrication costs, 83, 86, 91, 94
  in example quotation for CE, 83
  in final calculation for presentation for amount due for payment, 63
  in SCC, 53, 54–55, 63, 65, 83
  in SSCC, 58, 59, 63, 65, 96
materials
  Client not providing materials for inspection and tests, 13
  payments for use in construction or fabrication of Equipment, 53
  see also Plant and Materials
Met Office, 10, 11, 32
modifying Equipment as cost component
  SCC, 53
  SSCC, 58
multiple projects on same site, 64

NEC *see* New Engineering Contracts
New Engineering Contracts (NECs – family), 1
  NEC4, 2
    compared with NEC3, 101
notification of CEs, 20–22
  by *Contractor*, 21–22, 37, 38–39, 41
  by *Project Manager see* Project Manager
  flow charts, 38–41

offices, *Contractor's*, 103
omissions of work, 67–68
on-costs, Site, 101–104
on-Site requirements of *Client's* and *Consultant's*, 105
operatives as cost component
  SCC, 53
  SSCC, 58
Options
  main *see* main Options
  Option A, 24, 30, 44, 45, 45–46, 47, 48, 49, 50, 55, 56, 64, 66, 74, 85
  Option B, 5, 18, 24, 30, 44, 45, 46, 47, 48, 49, 56, 64, 66, 74, 95, 97
  Option C, 30, 46–47, 48, 49, 60, 63–64, 71, 73, 86, 108
  Option D, 44, 45, 46–47, 48, 49–50, 63–64, 71, 74, 95, 108
  Option E, 44, 45, 46–47, 48, 49–50, 63–64, 71, 74, 95
  Option F, 24, 48
  Option X2, 17, 92
  Option X10, 17
  Option X12, 17
  Option X14, 17, 47
  Option X15, 17
  Option Y(UK)2, 17
  Option Z, 7, 9, 10, 19
  secondary, 17, 22, 32
Others/other people, 7, 22
  defects corrected by, 68
  work to be done by, 7

payments, 59
  to *Contractor*, 45
    advanced, 54
    amount due, 17, 46, 47, 60–63, 64
    for Equipment in SCC, 52–53
    final calculation for presentation, 60–63
    final calculation and presentation, 60–63
    for materials used in construction or fabrication of Equipment, 53
  to people (*Contractor's* payroll), 88, 100
  to *Subcontractor*, 54, 65, 70, 82, 101
people/personnel/employees/staff
  costs (labour costs), 60, 69, 75, 77, 87–89, 95, 101, 102, 103
    in example quotations for CEs, 75–77, 95, 99–100
    in SCC, 51, 60, 75–77, 87–89, 102
    in SSCC, 56, 57, 60, 95, 102
    in supporting notes, 87–89
  directly employed by *Contractor*, 51, 54, 60, 65, 77
  expenses, 102
  People Rates, 57, 64, 69, 74, 89
  risk relating to, 91
  see also Others
performance, suspension, 17
personnel *see* people
physical conditions, 4, 8–9, 28, 31
plant
  Contractors's constructional plant, 102
  general purpose constructional plant, 102, 106
  not included in unit costs, 106
Plant and Materials, 53–54, 58–59, 81, 91
  Client not providing materials for inspection and tests, 13
  as cost component, 61, 81, 85, 86, 94
    in final calculation and presentation for amount due for payment, 61
    SCC, 53–54, 63, 65, 83, 105
    SSCC, 58–59, 65, 96, 105
  risk relating to, 91
preliminaries build-up, 70, 101–106
Price(s)
  changes to, 18–19, 23, 24–25
    assessment, 18–19, 23, 45, 66, 67
    reduction of, 28
Price for Work Done to Date, 17, 25, 45, 46, 47, 48, 64, 108
price(d)-based contracts, 5, 45–46, 48
  see also lump sum
programme
  Accepted Programme, 7, 13, 18, 23, 25, 26, 27
  in assessment in CEs, 27–28
Project Manager (*PM*), 36
  acceptance of CE by *see* quotations
  assessment of CE by, 26–28, 68
  instructions to submit quotation by, 20–22, 23, 26, 28, 36
  notification of CEs and, 20–21, 36
    failure of *PM* concerning notification of CE, 22, 30, 33
  flow chart, 38–39

notification by *PM*, 20–21, 36
roles (relating to CEs), 18
published list
  Equipment in, 57, 58, 61, 65, 74, 95
  Equipment not in, 58, 61

quality, risks relating to, 91
quotations for acceleration, 8, 23, 69
quotations for CEs, 23–27, 36, 37, 71–98
  acceptance, 26, 28, 36
  not accepted, 14, 20, 21, 68
  assessment, 26–27
  examples, 71–98
    SCC, 71–95
    SSCC, 95–96
  failure to submit, 23, 26, 42
  format, 33
  manual, 69, 89
  what to include, 23–24

rainfall, 11, 12
rates/Rates, 97, 103
  build up, 100
  People, 57, 64, 69, 74, 89
  unit, temporary works not included in, 105
reimbursement, 44, 46, 47, 48, 49, 50, 56
  additional, 9
  cost-reimbursable-type contracts, 31, 55
  direct, 50, 56
  indirect, 56
replying to communications, not, 7–8, 38
responsibilities *see* roles and responsibilities
risks, 90–93
  Client's, 92
  Contractor's, 90–93
    allocation/reallocation, 10, 31, 32, 93
    allowances, 90–93
    examples, 93, 96
    transferability, 31
roles and responsibilities (incl. duties) relating to CEs
  Clients *see* Clients
  Contractors, 18
  Project Managers, 18

safety requirements/matters, 7
  special, 92
salaries and wages, 88, 100, 102
samples, 81
  Client not providing them for inspection and tests, 13
Schedule of Cost Components (SCC), 43–70
  components of cost, 49–55
  example quotation for CE based on, 71–95
  final calculation and presentation for amount due for payment, 60–63
  practical issues, 64–70
  short form *see* Short Schedule of Cost Components
  summary of use, 48, 85, 87
  traditional preliminaries build-up compared with/related to, 101–108

understanding the role, 44
when to use, 45–46
Scope, 2
   CEs in, 6–7
Short Schedule of Cost Components (SSCC), 43–70, 101–108
   components of cost, 56–60
   example quotation for CE based on, 95–96
   final calculation and presentation for amount due for payment, 60–63
   practical issues, 64–70
   traditional preliminaries build-up compared with/related to, 101–108
   use, 64
      summary, 48
   when to use, 45–46
Site (and site)
   access *see* access
   Client's requirements on, 105
   Consultant's requirements on, 105
   multiple projects on same site, 64
   object of value or historical or other interest, 8, 20
   on-costs, 101–104
Site Information, 2, 9
   drawings, 108
staff *see* people
Subcontractors
   as cost component, 101
   in example quotation for CE, 82, 86
   in final calculation for presentation for amount due for payment, 61
   quotation for CE from *Subcontractor*, 86
   SCC, 54
   SSCC, 59
   fee percentage, 96, 108
   payments to, 54, 61, 65, 70, 82, 101
   risks *see* risks
   share, target cost contracts, 107–109
supporting notes in example quotation for CE, 87–95

take over of part of *works*, 13, 21
target cost contracts, 5, 46, 107–108
temporary works not included in unit rate, 106
tests, 8, 69
   Client not providing material/facilities/samples for, 13
time issues and timescales
   Client's and *Consultant's* requirement on Site, 102
   eight weeks since becoming aware of, 4, 21–22, 29

site on-costs, 102–103
stated in CE procedures, 19
submitting quotations, 23
time required for item of Equipment, 58
weather measurements, 10
see also dates
traditional contracts, 4, 10
   labour cost calculations, 100
   preliminaries build-up, 101–106
transport as cost component
   SCC, 53, 103
   SSCC, 58, 103

unit costs, plant not included in, 105
unit rates, temporary works not included in, 105

valuation *see* assessment

wages and salaries, 88, 100, 102
weather, 10–11, 32
   occurs on average less frequently than once in 10 years, 10, 11, 12
   weather data, 10, 11, 12
   weather measurement, 10, 11, 32
work (and *works*)
   already done, 18, 19, 24, 68, 87, 95
   completion of *see* Completion
   design outside, 55, 63, 73
   difference between final total quantity of work done and the quantity stated for an item in *Bill of Quantities*, 15–17
   efficient and effective management of, 4–5
   not yet done, 18, 24–25, 93
   omissions, 67–68
   payment for Equipment purchased for (included in contract), 52, 60, 73, 90
   Price for Work Done to Date, 17, 25, 45, 46, 47, 48, 64, 108
   section affected in example quotation for CE, 75–87
   take over of part of, 13, 21
   temporary, not included in unit rates, 105
   to be done by *Client* and *Others*, 7
Working Areas, 50, 101
   manufacture and fabrication outside, 54, 55, 56, 59, 63, 65, 73
   in SCC, 50, 51, 53, 54, 55, 61, 62, 63, 65, 73, 101
   in SSCC, 56, 57, 58, 59, 65, 73, 101
workmanship, risks relating to, 91